ASTROPHYSICS WITH A PC

An Introduction To Computational Astrophysics

Paul Hellings

Published by:

P.O. Box 35025
Richmond, Virginia 23235, USA
(804) 320-7016

Published by Willmann–Bell, Inc.
P.O. Box 35025, Richmond, Virginia 23235

First English Edition

Printed in the United States of America

Library of Congress Cataloging-in-Publication Data
Hellings, P. (Paul)
Astrophysics with a PC : an introduction to computational astrophysics / Paul Hellings. – 1st English ed.
p. cm.
Includes bibliographical references and index.
ISBN 943396-43-3
1. Astrophysics – Data processing. 2. Astrometry – Data processing. 3. Microcomputers. I. Title.
QB51.3.E43H45 1994
523'.01'0285'526 – dc20 94-30449
CIP

94 95 96 97 98 99 9 8 7 6 5 4 3 2 1

Foreword

Among all the branches of exact sciences, astronomy is certainly one of the best suited for popularization. Many important concepts can be explained and understood without having to use advanced mathematics. A small telescope is often sufficient to "discover" many details on the lunar surface or to observe the rings of Saturn, the large moons of Jupiter, solar spots, comets and planetoids, galactic star clusters and nebulae, and extragalactic systems. Thus, thousands of amateur astronomers have found an interesting and instructive hobby.

The explosive rise of the personal computer during recent years has provoked an important new activity among amateur astronomers. Since amateurs have a sufficient knowledge of positional astronomy concepts, such as coordinate systems, they are often able to write programs to compute ephemerides of all kinds without much difficulty. Locating celestial objects from the observer's own computations offers extra satisfaction. A large number of such programs are circulating among amateurs or may be bought for only a few dollars.

The subject of this book is something different: astrophysics. This discipline studies the internal structure and evolution of celestial bodies. Most amateurs are not too familiar with the concepts, terminology, and formulae used to describe astrophysical systems; therefore, we have decided to treat only a few cases in detail.

This book has two purposes: first, it offers every amateur an initiation into astrophysics; secondly, there is much to offer the PC owner. The book contains everything to allow the reader to write his own programs for computing a number of applications such as the internal structure of a star or a white dwarf, the dynamics of a universe model, the relevant processes of a stellar atmosphere, or the behavior of comets and meteoroids. Every subject is first introduced to illustrate its importance in the global framework of astronomy. Then, the relevant formulae and physical processes involved are discussed. As much as possible, these are justified by qualitative physical arguments, as an exact mathematical proof is beyond the purpose of this work. In the subsequent sections, we focus on the numerical expressions of these formulae and provide the reader with all the necessary information and, eventually, with flowcharts to allow him to write a correct program. Great care has been taken to include a maximum amount of detail. Finally, a number of examples and applications are given to enable the active reader to check the validity of his program.

A basic knowledge of mathematics and physics is needed to get the most out of the information presented. Therefore, it is absolutely necessary to study the first chapter, on numerical methods, before any of the other problems are attacked. The same methods are used as much as possible.

The examples and applications presented in this work have been programmed on personal computers (Olivetti M240, IBM PS/80, Macintosh SE and IIcx), and some on larger systems (PDP 11, CDC Cyber). However, any personal computer will do for all the problems presented. Results may differ slightly from one computer to another due to a different number of significant digits and rounding effects. The author wishes to thank the Groep T Institute of Technology (Louvain, Belgium) and the VUB (Free University of Brussels, Belgium) for the use of their computer facilities.

The author is indebted to Dr. Claude Doom and Mr. Jean Meeus for a critical reading of the manuscript and to Mr. Perry W. Remaklus of Willmann-Bell, Inc., for all his good advice and assistance in preparing this book for publication.

Paul Hellings, Ph.D.
June 1994

Table of Contents

Chapter 1

Some Numerical Methods

1.1 Introduction

Mathematical models describing physical processes or systems often contain differential equations and complicated integrals or require the solution of nonlinear equations or other typical problems. Since most of such mathematical applications cannot be solved exactly, or to use the correct word "analytically," we have to work with all kinds of approximations or successive iterations. A well-known example in positional astronomy is the solution of Kepler's equation for elliptic motion. The problem is to compute the eccentric anomaly E from the nonlinear equation

$$E = M + e \sin E$$

where M is the given mean anomaly and e the given eccentricity of the orbit. This equation is important when computing the position of a small body in a central gravitational field, such as a planet in its orbit around the Sun. It is not possible to solve this equation analytically, i.e., to express E as a function of M and e. It is, however, possible to obtain the solution E by successive approximations using the recursion formula

$$E_{i+1} = M + e \sin E_i.$$

Starting from the initial value $E_0 = M$, one puts at every iteration the last computed value of E in the right-hand side to obtain a new approximation. The list of successive approximations converges to the exact solution. The number of iterations needed is dependent on the accuracy demanded by the user and on the magnitude of the eccentricity e.

Similar techniques of approximation by iteration exist for all kinds of mathematical problems. Through numerous simple mathematical calculations these techniques transform a mathematically difficult problem into an algorithm. Thus, these techniques are especially suited for application on a computer. In the following sections of this chapter we will discuss some iterative methods. The branch of mathematics dealing with these methods is *numerical analysis*. It is perhaps the most frequently used branch of mathematics in scientific applications.

The fact that we deal with approximations and computers also means that we make errors. They may arise from the limited number of significant digits in a computer (rounding error) and from the limited number of iterations (truncation error). The evaluation of these errors often requires additional and mostly complicated computations. We have decided not to include such evaluations in our programs. Whenever possible, we will compare our results with professional data in order to estimate the validity of our often simplified models. We refer to textbooks on numerical analysis for a profound error analysis.

1.2 Numerical Solutions of Differential Equations

Differential equations are equations in which the unknown is not a variable x, but a complete function $y(x)$. The equation may contain not only the unknown function $y(x)$ and eventually the independent variable x, but also the derivative $y'(x)$, $y''(x)$, and so on. These derivatives are sometimes written dy/dx, $d^2y/dx^2 \ldots$ for the first and second derivative. Some examples of differential equations with their general solutions are as follows:

Equation	General Solution
$y' = -2xy$	$y = A\exp(-x^2)$
$y'' = -y$	$y = A\cos(x) + B\sin(x)$
$y'' - 4y' + 3y = 0$	$y = A\exp(x) + B\exp(3x)$
$y'' - 4y' + 3y = x^2$	$y = A\exp(x) + B\exp(3x) + \frac{26}{27} + \frac{8}{9}x + \frac{1}{3}x^2$
$y''' + 2y'' - y' - 2y = 0$	$y = A\exp(x) + B\exp(-x) + C\exp(-2x)$

The factors A, B, $C\ldots$ in the general solutions are constants that may be chosen freely. Every choice selects another solution of the differential equation. This means that the solution $y(x)$ of a differential equation is not a unique function but a complete set of functions, all satisfying the differential equations and each one of them characterized by a specific choice of the free constant(s) A, B, $C\ldots$. Furthermore, we see that the number of free constants is equal to the degree of the highest derivative, one in the first example, two in the second to the fourth, and three in the last. The free constants (called integration constants) may be used to select the solution that satisfies an additional condition, usually the passage of the solution through a given point (x_0,y_0). The number of such additional conditions that one may demand is equal to the number of free constants.

If we demand, for instance, the solution of the first example passing through the point ($x = 2$, $y = 3$), we simply introduce this condition in the general solution and therefore obtain the correct value of A:

$$3 = A\exp(-4)$$

thus

$$A = 163.794450\ldots.$$

For the second we could demand the passage through two points, or the passage through one point, with a given value of the first derivative at that point. For instance, what is the solution through (0,2) having $y' = -1$ in $x = 0$?

First condition:

$$2 = A\cos(0) + B\sin(0)$$

thus

$$A = 2.$$

Second condition:

$$-1 = -A\sin(0) + B\cos(0)$$

thus

$$B = -1.$$

The unique solution satisfying these two conditions is

$$y(x) = 2\cos(x) - \sin(x)$$

which no longer contains any integration constant. Such additional conditions are called *boundary* or *initial* conditions. A differential equation provided with the right number of boundary conditions is called a *differential problem*.

Unfortunately, most of the differential equations may not be solved directly. It is not possible to find and write down a certain function $y(x)$ satisfying the problem. Therefore, we have to solve the problem numerically by means of successive iterations. In the next sections we will present some basic techniques of the numerical solution of a differential equation. We will first focus on single first-order equations. Once these are mastered, it is easily possible to consider systems of such equations or to reduce certain second-order equations to a system of first-order equations.

The first-order equations we will consider all have the form

$$y' = f(x, y).$$

The purpose of this calculation is to find the solution starting from a given point (x_0, y_0) by successive iterations.

1.2.1 Euler's Method

The first method is the most simple one, but also the least accurate. The solution $y(x)$ in the vicinity of a point (x_i,y_i) on the solution is approximated along the tangential T in that point. Suppose we have a point (x_i,y_i) of the solution. Let us then proceed to x_{i+1}, a new x-point a little bit (dx) further. Thus

$$x_{i+1} = x_i + dx. \tag{1}$$

Since the slope of the tangential T to the solution in (x_i,y_i) is given by the value of the derivative y' in that point, the equation of T is

$$y - y_i = y'(x_i)(x - x_i).$$

But since $y' = f(x,y)$

$$y - y_i = f(x_i, y_i)(x - x_i)$$

So if we put x equal to x_{i+1}, using (1) we obtain

$$y_{i+1} = y_i + dx f(x_i, y_i) \tag{2}$$

where y_{i+1} is an approximation of the y-value in x_{i+1}. This point $(i+1)$ is then used to go once again dx further to a point $(i+2)$. By repeating this calculation a number of times starting from the given initial point (x_0,y_0), an approximation for the solution of the differential problem can be found.

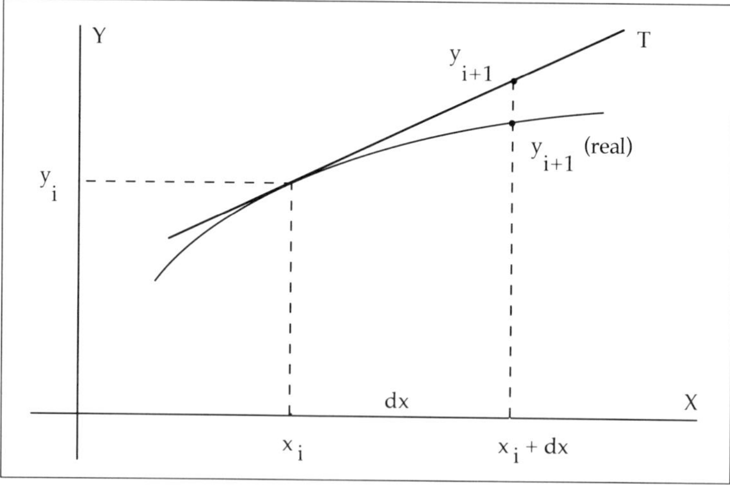

Fig. 1.1: *Euler's method.*

Each time we compute a new point we make a small error. This is demonstrated in Fig. 1.1 as the difference between the computed point $(i+1)$ and the exact (but unknown) point on the real solution. Euler's method does not give the exact solution since it approximates a tangent only over a finite distance. The method behaves like a car entering a curve in the road with a speed that is too high. The error becomes smaller when dx is decreased, but in that case a larger number of iterations is needed to obtain a solution over an interval of a fixed length, and rounding errors become relatively larger. The choice of an appropriate step-size dx is important. We will suggest suitable values for the step-size in all our applications throughout this book.

The following sample program is written in MicroSoft QuickBasic 4.5:

```
DECLARE FUNCTION f (x, y)

'
'  Sample program to solve the differential equation
'
'                    y' = f(x,y)
'
'  using Euler's method
'
'  The exact form of the right hand side is placed in FUNCTION F
'  of this program

INPUT "Initial condition   x(0) = "; x
INPUT "                    y(0) = "; y
INPUT "Stepsize            dx   = "; dx
INPUT "Number of iteration to be computed = "; n%

FOR i% = 1 TO n%

'  compute new iteration
    x1 = x + dx
    y1 = y + dx * f(x, y)

'  display new iteration
    PRINT USING "####      #####.#####       #####.#######"; i; x1; y1

'  prepare for next iteration
    x = x1
    y = y1

NEXT i%


FUNCTION f (x, y)
'
'  contains the right hand side of the differential equation
'
    f = -2 * y * x

END FUNCTION
```

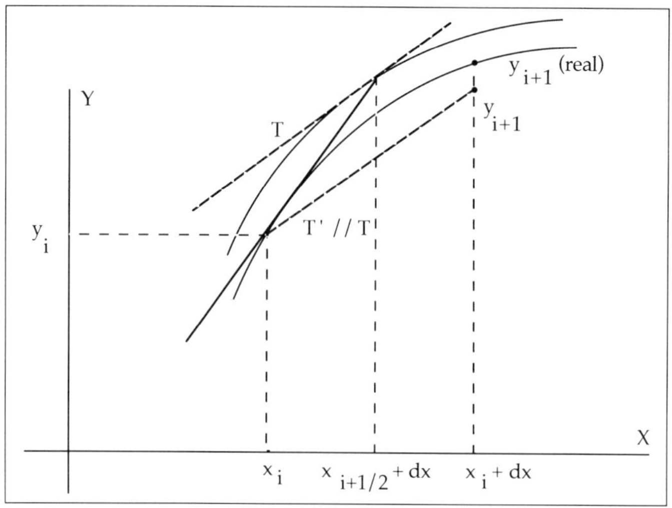

Fig. 1.2: *Midpoint method performing the step i to $i+1$ but with the slope at half the step in $i+1/2$.*

1.2.2 Midpoint Method (Cauchy Method)

This method uses the slope of the tangent at half the step. It is performed by applying Euler's method with a step $(1/2)dx$. At that point $(x_{i+1/2}, y_{i+1/2})$ the slope of the tangential is then computed. With this slope we return to (x_i, y_i) to perform the whole step over dx but with the slope of the point $i+1/2$.
First compute an approximation for the solution at $x_{i+1/2}$ from these formulae:

$$x_{i+1/2} = x_i + 0.5dx \tag{3}$$

$$y_{i+1/2} = y_i + 0.5dxf(x_i, y_i). \tag{4}$$

This is nothing more than Euler's method performed with a step-size $0.5dx$. The slope of the tangential in $(x_{i+1/2}, y_{i+1/2})$ is the value of the derivative; therefore $f(x_{i+1/2}, y_{i+1/2})$.

We now perform with this slope the whole step over dx from x_i to x_{i+1}:

$$x_{i+1} = x_i + dx \tag{5}$$

$$y_{i+1} = y_i + dxf(x_{i+1/2}, y_{i+1/2}). \tag{6}$$

Equations (3–6) form the midpoint method to go from point i to $i+1$. This method is more accurate than Euler's method. However, the number of calculations needed to perform one step is twice Euler's method since we need to evaluate $f(x, y)$ two times, once in point (x_i, y_i) and a second time in $(x_{i+1/2}, y_{i+1/2})$.

The following sample program is written in MicroSoft QuickBasic 4.5:

```
DECLARE FUNCTION f (x, y)

'
'  Sample program to solve the differential equation
'
'                    y' = f(x,y)
'
'  using the midpoint method
'
'  The exact form of the right hand side is placed in FUNCTION F
'  of this program


INPUT "Initial condition   x(0) = "; x
INPUT "                    y(0) = "; y
INPUT "Stepsize            dx   = "; dx
INPUT "Number of iteration to be computed = "; n%

FOR i% = 1 TO n%

'  compute first half of new iteration
    x12 = x + .5 * dx
    y12 = y + .5 * dx * f(x, y)

'  compute complete iteration with slope at half the step
    x1 = x + dx
    y1 = y + dx * f(x12, y12)

'  display new iteration
    PRINT USING "####      #####.#####       #####.#######"; i; x1; y1

'  prepare for next iteration
    x = x1
    y = y1

NEXT i%


FUNCTION f (x, y)
'
'  contains the right hand side of the differential equation
'
    f = -2 * y * x

END FUNCTION
```

1.2.3 Predictor-corrector Method (Heun Method)

This method has the same accuracy and number of calculations as the midpoint method. The slope used now is the average of the slope in point (x_i, y_i) and in point (x_{i+1}, y_{i+1}), the latter obtained with Euler. In fact Euler's method is used as a preliminary solution, which is then corrected to obtain the final result, as follows:

For the predictor,

$$x_{i+1} = x_i + dx \tag{7}$$

$$y_{i+1}^P = y_i + dx f(x_i, y_i). \tag{8}$$

For the corrector,

$$y_{i+1} = y_i + 0.5dx\{f(x_i, y_i) + f(x_{i+1}, y_{i+1}^P)\}. \tag{9}$$

It is easy to see that the function $f(x, y)$ has to be evaluated two times, just as for the midpoint method.

The following sample program is written in MicroSoft QuickBasic 4.5:

```
DECLARE FUNCTION f (x, y)

'
'  Sample program to solve the differential equation
'
'                      y' = f(x,y)
'
'  using the predictor-corrector method
'
'  The exact form of the right hand side is placed in FUNCTION F
'  of this program


INPUT "Initial condition    x(0) = "; x
INPUT "                     y(0) = "; y
INPUT "Stepsize             dx   = "; dx
INPUT "Number of iteration to be computed = "; n%

FOR i% = 1 TO n%

'  compute the predicted new iteration
     x1p = x + dx
     y1p = y + dx * f(x, y)

'  compute the corrected new iteration
     x1 = x + dx
     y1 = y + .5 * dx * (f(x, y) + f(x1p, y1p))

'  display new iteration
     PRINT USING "####      #####.#####        #####.#######"; i; x1; y1

'  prepare for next iteration
     x = x1
     y = y1

NEXT i%


FUNCTION f (x, y)
'
'  contains the right hand side of the differential equation
'
    f = -2 * y * x

END FUNCTION
```

1.2.4 Runge-Kutta Method of the Fourth-Order

This method is superior to the other three, but it needs four times the computation of $f(x, y)$. It uses a complicated combination of four slopes of tangentials over the interval dx. The method gives excellent results but is rather extended to use in small programs, especially if not one but a large system of differential equations has to be solved. We have therefore decided to work whenever possible with the midpoint method throughout this monograph. In this way it is possible to solve the problems with sufficient accuracy and in a reasonable length of time. Experienced readers owning a PC may of course adapt the numerical descriptions to the Runge-Kutta method. The Runge-Kutta method is as follows:

$$x_{i+1/2} = x_i + 0.5dx \tag{10}$$

$$x_{i+1} = x_i + dx \tag{11}$$

Thus,

$$K1 = dxf(x_i, y_i) \tag{12}$$

$$K2 = dxf\left(x_{i+1/2}, y_i + 0.5K1\right) \tag{13}$$

$$K3 = dxf\left(x_{i+1/2}, y_i + 0.5K2\right) \tag{14}$$

$$K4 = dxf(x_{i+1}, y_i + K3) \tag{15}$$

$$y_{i+1} = y_i + \frac{K1 + 2K2 + 2K3 + K4}{6}. \tag{16}$$

The following sample program is written in MicroSoft QuickBasic 4.5:

```
DECLARE FUNCTION f (x, y)

'
'  Sample program to solve the differential equation
'
'                        y' = f(x,y)
'
'  using the Runge-Kutta method
'
'  The exact form of the right hand side is placed in FUNCTION F
'  of this program


INPUT "Initial condition    x(0) = "; x
INPUT "                     y(0) = "; y
INPUT "Stepsize             dx   = "; dx
```

```
INPUT "Number of iteration to be computed = "; n%

FOR i% = 1 TO n%
'
    x1 = x + dx

    k1 = dx * f(x, y)
    k2 = dx * f(x + .5 * dx, y + .5 * k1)
    k3 = dx * f(x + .5 * dx, y + .5 * k2)
    k4 = dx * f(x + dx, y + k3)

    y1 = y + (k1 + 2 * k2 + 2 * k3 + k4) / 6

'   display new iteration
    PRINT USING "####      #####.#####       #####.#######"; i; x1; y1

'   prepare for next iteration
    x = x1
    y = y1

NEXT i%


FUNCTION f (x, y)
'
'   contains the right hand side of the differential equation
'
    f = -2 * y * x

END FUNCTION
```

1.2.5 Some Examples

In this subsection are approximations of a differential problem with a known, exact solution. In this way, we are able to compare the results obtained with the four numerical methods described above. The problem is

$$y' = -2xy$$

with $y = 1$ at $x = 0$.

We start from $x_0 = 0$ and $y_0 = 1$ and use a step-size $dx = 0.1$ in all the methods. The exact solution is

$$y = \exp(-x^2).$$

The results are given in the following Table 1-1. The reader is invited to program and reproduce the results of some of these methods as a first exercise in applying numerical methods for the solution of differential equations. Special attention should be paid to the midpoint method since this method will be applied in most of the other chapters.

1.2.6 System of Simultaneous First-Order Equations

We have discussed in the previous subsections some numerical methods to solve a first-order equation $y' = f(x, y)$ with initial conditions $y_0 = y(x_0)$. In most applications, however, the mathematical description of the physical

Table 1-1.

x	y(Euler)	y(Mid pt.)	y(pred. cor)	y(Runge K)	y(exact)
0.00	1.00000000	1.00000000	1.00000000	1.00000000	1.00000000
0.10	1.00000000	0.99000000	0.99000000	0.99004983	0.99004983
0.20	0.98000000	0.96059700	0.96069600	0.96078944	0.96078944
0.30	0.94080000	0.91352775	0.91381404	0.91393117	0.91393118
0.40	0.88435200	0.85149921	0.85204021	0.85214377	0.85214379
0.50	0.81360384	0.77792968	0.77876475	0.77880078	0.77880078
0.60	0.73224346	0.69663603	0.69777321	0.69767639	0.69767633
0.70	0.64437424	0.61150711	0.61292399	0.61262661	0.61262639
.	.	.	.	.	.
.	.	.	.	.	.
.	.	.	.	.	.
1.60	0.06751343	0.07812552	0.07929913	0.07731173	0.07730474
1.70	0.04590914	0.05646913	0.05744429	0.05558359	0.05557621
1.80	0.03030003	0.04006485	0.04085438	0.03917137	0.03916390
.	.	.	.	.	.

model contains a number of simultaneous or coupled first-order equations. In such a system we have to solve a number of differential equations, all having their own initial condition. These equations should be solved simultaneously, since the right hand side of each differential equation may contain a number or even all of the other functions. The general form is as follows:

$$y' = f_y(x, y, z, u \ldots)$$

with y_0 given at x_0;

$$z' = f_z(x, y, z, u \ldots)$$

with z_0 given at x_0;

$$u' = f_u(x, y, z, u \ldots)$$

with u_0 given at x_0.

The total order of the system is equal to the number of equations since each equation is first-order. Following is an example of such a system of two coupled differential equations:

$$y' = -2z$$

with $y(x_0 = 0) = 1$, and

$$z' = y - \sin^2 x$$

with $z(x_0 = 0) = 0$.

In this example, $f_y = -2z$ and $f_z = y - \sin^2 x$. The problem is to find the solution for both the unknown functions $y(x)$ and $z(x)$. Finding analytical solutions is almost never possible. Once again, such systems are mostly solved by numerical iterations.

The numerical solution of such a system is quite easy when we apply one of the methods given for single equations. We will demonstrate this with the midpoint method. Similar techniques are applied for the other methods such as Euler, predictor-corrector, or Runge-Kutta. Suppose we have the values of $y, z, u, \ldots$ at a certain point x_i. Let us call these values $y_i, z_i, u_i, \ldots$. To obtain the values at x_{i+1}, i.e., $y_{i+1}, z_{i+1}, u_{i+1}, \ldots$, for a step-size dx, we first compute the values at half the step as follows:

$$x_{i+1/2} = x_i + \frac{1}{2}dx \tag{17}$$

$$y_{i+1/2} = y_i + \frac{1}{2}dx f_y(x_i, y_i, z_i, u_i, \ldots) \tag{18}$$

$$z_{i+1/2} = z_i + \frac{1}{2}dx f_z(x_i, y_i, z_i, u_i, \ldots) \tag{19}$$

$$u_{i+1/2} = u_i + \frac{1}{2}dx f_u(x_i, y_i, z_i, u_i, \ldots) \tag{20}$$

$$\ldots$$

The whole step is computed as follows:

$$x_{i+1} = x_i + dx \tag{21}$$

$$y_{i+1} = y_i + dx f_y(x_{i+1/2}, y_{i+1/2}, z_{i+1/2}, u_{i+1/2}, \ldots) \tag{22}$$

$$z_{i+1} = z_i + dx f_z(x_{i+1/2}, y_{i+1/2}, z_{i+1/2}, u_{i+1/2}, \ldots) \tag{23}$$

$$u_{i+1} = u_i + dx f_u(x_{i+1/2}, y_{i+1/2}, z_{i+1/2}, u_{i+1/2}, \ldots) \tag{24}$$

$$\ldots$$

These formulae show that it is necessary to compute all the functions in half-steps before one of them can be evaluated in $i+1$. This is the simultaneous or coupled aspect of the method. A system of two coupled equations is more than two separate equations: either you solve them both, or you don't solve anything at all. Let us return to the example:

$$y' = -2z$$

with $y(0) = 1$, and

$$z' = y - \sin^2 x$$

with $z(0) = 0$.

The exact solution of this problem is

$$\begin{aligned} y(x) &= \cos^2 x \\ z(x) &= \sin x \cos x. \end{aligned}$$

We will compute numerical approximations of this solution with the formulae given above (omit u!) and compare the results with the exact solution. Using a step-size $dx = 0.05$ in equations (17, 18, 19, 21, 22, 23) the following values are obtained:

x	y	y (exact)	z	z (exact)
0.00	1.000000	1.000000	0.000000	0.000000
0.05	0.997500	0.997502	0.049969	0.049917
0.10	0.990015	0.990033	0.099438	0.099335
0.15	0.977621	0.977668	0.147913	0.147760
0.20	0.960442	0.960530	0.194908	0.194709
0.25	0.938648	0.938791	0.239954	0.239713
0.30	0.912459	0.912668	0.282600	0.282321
0.35	0.882136	0.882421	0.322418	0.322109
0.40	0.847983	0.848353	0.359012	0.358678
0.45	0.810341	0.810805	0.392013	0.391663
0.50	0.769587	0.770151	0.421092	0.420735
.	.	.	.	.
.	.	.	.	.
.	.	.	.	.

The following sample program is written in MicroSoft QuickBasic 4.5:

```
DECLARE FUNCTION fy (x, y, z)
DECLARE FUNCTION fz (x, y, z)


'
'  Sample program to solve the system of two coupled differential equations
'
'                    y' = fy(x,y,z)
'                    z' = fz(x,y,z)
'
'  using the simultaneous midpoint method
'
'  The exact form of the right hand sides are placed in FUNCTION FY
'  and FUNCTION FZ of this program


INPUT "Initial condition   x(0) = "; x
INPUT "                    y(0) = "; y
INPUT "                    z(0) = "; z
INPUT "Stepsize            dx   = "; dx
INPUT "Number of iteration to be computed = "; n%

FOR i% = 1 TO n%

'  compute first half of new iteration
    x12 = x + .5 * dx
    y12 = y + .5 * dx * fy(x, y, z)
    z12 = z + .5 * dx * fz(x, y, z)

'  compute full iteration with slope at half the step
    x1 = x + dx
    y1 = y + dx * fy(x12, y12, z12)
    z1 = z + dx * fz(x12, y12, z12)

'  display new iteration
    PRINT USING "####   #####.#####   #####.######   #####.######"; i; x1; y1; z1

'  prepare for next iteration
    x = x1
    y = y1
    z = z1

NEXT i%


FUNCTION fy (x, y, z)
'
'  contains the right hand side of the differential equation
'  y' = fy(x,y,z)
'
    fy = -2 * z

END FUNCTION


FUNCTION fz (x, y, z)
'
'  contains the right hand side of the differential equation
'  z' = fy(x,y,z)
'
    fz = y - (SIN(x)) ^ 2

END FUNCTION
```

1.2.7 Reducing a Second-Order Equation into Two First-Order Equations

Most of the mechanical models involve the acceleration of a particle as a consequence of the forces affecting it. The acceleration is the second derivative of the position $r(t)$, and the velocity is the first, leading to second-order equations of the form

$$y'' = f(x, y, y')$$

with y_0 and y'_0 given at some initial point x_0.

The independent variable x is then interpreted as the time. We have to specify two initial conditions, one for the initial position and one for the initial velocity. Such a problem is easily transformed into two first-order equations by considering the first derivative y' as a separate function $z(x)$. Thus, we have

$$y' = z \tag{25}$$

with $y(0) = y_0$. Since $y' = z$, we also have $y'' = z'$:

$$z' = f(x, y, z) \tag{26}$$

with $z(0) = z_0 = y'_0$.

This new problem is then solved by the techniques described in section 2.6 since it is now a system of two coupled first-order equations with given initial conditions.

An example in which such a transformation is used is simply the motion in a central gravitational field in two dimensions. According to Newton's law the acceleration is given by

$$a(t) = -G\frac{M}{r^2(t)}.$$

For the acceleration in the x and y direction the formula is

$$x'' = -G\frac{M}{x^2 + y^2}\cos h \quad y'' = -G\frac{M}{x^2 + y^2}\sin h.$$

After elimination of the angle h this becomes

$$x'' = -G\frac{M}{(x^2 + y^2)^{3/2}}x \quad y'' = -G\frac{M}{(x^2 + y^2)^{3/2}}y.$$

Each of these two second-order equations is then reduced to two first-order equations by introducing the two velocity components u and v for the x and

y directions as a separate function of the time. Furthermore, since $x' = u$, we also have $x'' = u'$ and the same for $y'' = v'$. The system then becomes

$$x' = u$$

$$y' = v$$

$$u' = -G\frac{M}{(x^2 + y^2)^{3/2}}x$$

$$v' = -G\frac{M}{(x^2 + y^2)^{3/2}}y.$$

Thus, the gravitational law of Newton in two dimensions is in fact a system of four coupled first-order differential equations. The unknown functions are the two components of the position (x, y) and the two components of the velocity (u, v). The independent variable is the time t. To solve this system numerically, we have to select four initial conditions: the initial position (x_0, y_0) and the initial velocity (u_0, v_0). The role of x in the general equations of section 2.6 is played here by the time t. The same problem in three dimensions is described by a system of three second-order equations, in which case the total order becomes six.

1.3 Computing Zeros of Functions

In this section two methods to locate a zero x_0 of a nonlinear function $y = f(x)$ will be discussed. It is assumed that all mathematical conditions concerning the continuity and differentiability are satisfied. We will also mention some typical problems in applying these methods. Such problems often involve the selection of good initial values. Much attention is normally paid to this selection since it may affect not only the speed of convergence, but also whether there is any convergence in the first place! Since our attention is mainly focused on the astrophysical aspects, we will provide the reader with suitable initial conditions each time we use one of these methods in one of our astrophysical applications so that these numerical problems are avoided.

1.3.1 Newton-Raphson Method

This numerical method may be explained in various ways, but the geometrical way is the most clear. Suppose we want to compute the zero x_0 of a function $y = f(x)$ and we know more or less where that zero lies. This computation is normally possible by making a quick sketch of the function.

If x_i is such a first approximation of the zero, a better approximation x_{i+1} may then be found in the following way. We draw the tangential T to the

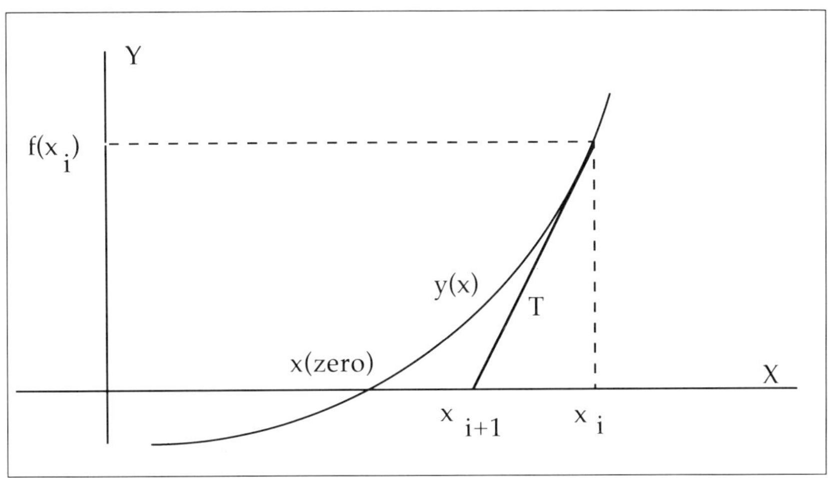

Fig. 1.3: *The Newton-Raphson method.*

function in $(x_i, f(x_i))$. The zero of that tangential intersects the X-axis in a certain point, and that point is then a better approximation x_{i+1} for the zero of the function.

This method, which is demonstrated in Fig. 1.3, is valid if the function "behaves" well and if the first approximation is close enough to the zero. Mathematically, the equation of the tangent line is

$$y - f(x_i) = f'(x_i)(x - x_i). \tag{27}$$

Its zero is obtained by setting $y = 0$. This zero is then called x_{i+1}, a better approximation. It is found through the Newton-Raphson formula, as follows:

$$x_{i+1} = x_i - \frac{f(x_i)}{f'(x_i)}. \tag{28}$$

A series of successive approximations, which began with a good initial guess, converges under certain conditions to the nearest zero of the function.

- *Example:* $f(x) = \exp(x) - 6x$.

This function has two zeros, one near 3 and one near 0. Starting from an initial guess $x = 0.0$, the successive approximations are as follows:

$$x1 = 0.200000000$$
$$x2 = 0.204478879$$
$$x3 = 0.204481449$$
$$x4 = 0.204481449$$

The last approximation is equal (at least for this number of digits) to the one just before it; therefore, it may be considered a good approximation for the exact zero.

Starting from another initial guess $x = 3.0$, one finds the following series of approximations:

$$x1 = 2.85193771$$
$$x2 = 2.83341629$$
$$x3 = 2.83314795$$
$$x4 = 2.83314789$$
$$x5 = 2.83314789$$

Again, the last approximation is equal to the one preceding it, so this is considered to be the second zero of that function.

Following are some typical problems with this method:

- An approximation lies outside the definition region of the function. This may occur in some cases during the approximation.
- The derivative of the function in the zero (or in a region near the zero) is zero itself (or very small). This may increase the number of approximations considerably or even prevent the method from converging. This problem may occur when the slope in one of the iteration points is very flat. In that case, the zero of the tangential in that point may be situated at a very large distance from the previous approximation.
- The function has more than one zero. Sometimes the method continually converges to the same zero, no matter which initial guess is selected. If this is the zero needed there is no problem; otherwise, another method must to used.

The following sample program is written in MicroSoft QuickBasic 4.5:

```
DECLARE FUNCTION f (x)
DECLARE FUNCTION dfdx (x)

'
' Sample program to search a zero of a non linear function f(x)
'
' The function is placed in FUNCTION f, and its first derivative
' in FUNCTION dfdx
'
' Input for this program are :
'
'   An initial guess in the neighborhood of the solution
'   to start the iterations
'
'   A maximum number of iterations allowed
'
'   A small number epsilon (for instance 0.0001), expressing the criterion
'   of convergence. When two successive iterations are closer to each other
```

```
'   than epsilon, the last iteration is considered as the correct solution
'
'
INPUT "Initial guess of the zero              : "; x
INPUT "Maximum number of iterations allowed : "; maxiter%
INPUT "Convergence criterion epsilon          : "; epsilon

iter% = 0


DO

'    save previous approximation for use in convergence test
     xold = x

'    compute new approximation
     x = x - f(x) / dfdx(x)

'    update number of iterations
     iter% = iter% + 1

     PRINT USING "###    #####.#########"; iter%; x

LOOP UNTIL (iter% > maxiter%) OR (ABS(x - xold) < epsilon)


IF iter% > maxiter% THEN
    PRINT "Maximum number of iterations used"
ELSE
    PRINT "method has converged: x = "; x
END IF


END



FUNCTION dfdx (x)
'
' contains the first derivative of the function f(x)
'

   dfdx = EXP(x) - 6

END FUNCTION



FUNCTION f (x)
'
'  contains the function for which a zero is searched for
'
     f = EXP(x) - 6 * x

END FUNCTION
```

1.3.2 Fixpoint Method

In this method, the solution of $f(x) = 0$ is replaced by the solution of a problem

$$x = g(x) \tag{29}$$

which is, for instance, obtained by eliminating a factor x from the function

$f(x)$. The iterations to be performed in that case are

$$x_{i+1} = g(x_i). \tag{30}$$

A graphical interpretation of this method may be found in every textbook on numerical analysis.

- *Example:* Find the zero(s) of $f(x) = x^2 - 3x - 4$.

This function has two zeros: $x = 4$ and $x = -1$. A first way to write down the iteration procedure is to eliminate the linear x factor

$$x = \frac{1}{3}(x^2 - 4)$$

so that the iterations are computed from

$$x_{i+1} = \frac{1}{3}(x_i^2 - 4).$$

Starting from an initial value selected between -4 and $+4$ always gives us the zero $x = -1$. When we start from exactly -4 or $+4$ we immediately find the other zero $x = 4$. Every initial value smaller than -4 or larger than $+4$ results in a divergent series. There is no possible solution.

Another way is to eliminate the x-factor with the square

$$x = \sqrt{3x + 4}.$$

All approximation series started from values larger than $-(4/3)$ lead to the solution $x = 4$. The other zero $x = -1$ cannot be reached. Initial guesses smaller than $-(4/3)$ cannot be used since they would require the computation of the square root of a negative quantity.

This example shows that it is very important to use a good iteration formula and to start from suitable initial values. Otherwise, the method may converge to another zero than the one looked for, or convergence may even be lost. The method is therefore only applied after a careful study of the function.

1.4 Numerical Computation of Integrals

The method described in this section is applied on a finite interval with lower boundary a and upper boundary b. The problem is to compute an approximation for the integral of a "difficult" function $f(x)$ over that interval. Thus

$$I = \int_a^b f(x)dx. \tag{31}$$

This integral gives the surface of the area S between the function $f(x)$ and the X-axis between the two boundaries. We assume that $f(x)$ is positive over the interval. The result of the integral is

$$F(b) - F(a)$$

where F is a primitive function of $f(x)$, i.e., a function for which

$$F'(x) = f(x).$$

In a number of practical cases, $f(x)$ is too complicated to obtain $F(x)$ directly; therefore, the surface S has to be computed numerically by a certain approximation. Some of these methods compute such an approximation directly, while others try to evaluate the integral within a certain accuracy selected by the user. Of course, methods of this second type need a greater number of calculations.

1.4.1 Fitting a Parabola to Obtain a Crude Estimate

This method is based on the function values $f(a)$, $f(b)$ and the function value in the middle of the interval $(a+b)/2$. Let us assume that the interval is centered around zero, so that $a = -b$. The coordinates of a and b are then, for instance, $x(a) = -h$, $x(b) = h$, $x(\text{middle}) = 0$. The corresponding values of $f(x)$ are then for the lower boundary a, $f(-h)$; for the upper boundary b, $f(h)$; and for the middle, $f(0)$. Through the three points $(-h, f(-h))$, $(0, f(0))$, and $(h, f(h))$, we may then consider a unique parabola

$$y = A + Bx + Cx^2$$

which has to satisfy

$$\begin{aligned} f(-h) &= A - Bh + Ch^2 \\ f(0) &= A \\ f(h) &= A + Bh + Ch^2, \end{aligned}$$

from which

$$\begin{aligned} A &= f(0) \\ B &= \frac{f(h) + f(-h)}{2h} \\ C &= \frac{f(-h) - 2f(0) + f(h)}{h^2}. \end{aligned}$$

If we agree to consider the parabola to be an approximation for the function $f(x)$, the same holds for their integrals: the surface of the area S' under the parabola may be considered as an approximation for the surface of the area S under $f(x)$. This area under the parabola is given by the integral of the parabola between $-h$ and h and is easily found, since it is an integral of powers of x. The result is

$$S = h\frac{f(-h) + 4f(0) + f(h)}{3}. \tag{32}$$

In this way we have a general rule to approximate the integral of any function $f(x)$ by using only the function values at the boundaries and in the middle of the interval, and the step-size $h = (a+b)/2$. It is obvious that this approximation will be very crude, since we use only three points to simulate the shape of the function. A possible way to improve the method would be to use higher order polynomials through more mesh points over the total interval of integration, but there is a more effective way still using parabolas.

1.4.2 Simpson's Method

The parabola method is only a first approximation. In this second method the parabola method is applied on a number of subintervals covering the global interval $[a, b]$. Let us divide the interval $[a, b]$ in an *even* number n of subintervals, all having the same size:

$$h = \frac{b-a}{n}.$$

The mesh points on the X-axis are

$$a = x_0,\ x_1 = x_0 + h,\ x_2 = x_0 + 2h, \ldots$$
$$\ldots x_{n-1} = x_0 + (n-1)h,\ x_n = x_0 + nh = b.$$

When we start with the first two subintervals, the integral of the function between x_2 and x_0 may be approximated using the parabola method

$$I(0,2) = h\frac{f_0 + 4f_1 + f_2}{3}.$$

Similarly, for the next two subintervals we have

$$I(2,4) = h\frac{f_2 + 4f_3 + f_4}{3}$$

and so on. For the last two subintervals we use

$$I(n-2,n) = h\frac{f_{n-2} + 4f_{n-1} + f_n}{3}.$$

If all these contributions of two successive subintervals are added, we finally obtain the approximation for the integral of the function between the boundaries a and b:

$$I = h\frac{f_0 + 4f_1 + 2f_2 + 4f_3 + 2f_4 + 4f_5 \ldots + 2f_{n-2} + 4f_{n-1} + f_n}{3}. \tag{33}$$

This is Simpson's method. Meshpoints with an uneven number have their function value multiplied with a factor 4, the other with a factor 2 except for the first (number 0) and the last (number n). The whole summation is multiplied with the step-size h and divided by 3.

- *Example:* Compute the integral of $f(x) = \exp(-x) + 2x$ over (2,6).

A primitive function of this $f(x)$ is easily found by

$$F(x) = -\exp(-x) + x^2.$$

The integral is $F(6) - F(2) = 32.13285653$.

When computed with various values of n we obtain the following data:

$n = 2$: 32.14071773	$n = 20$: 32.13285771
$n = 4$: 32.13351513	$n = 50$: 32.13285656
$n = 8$: 32.13290133	$n = 100$: 32.13285654

Try to compute a few values as an exercise.

1.4.3 How to Program Simpson's Method

We have seen in the previous subsection that the interval $[a, b]$ is divided in an even number of n successive subintervals. We then consider $n/2$ times a pair of two adjacent subintervals. For each pair a quantity of the parabola formula has to be computed. It is obvious that this formula should be programmed in a `FOR` loop, which is executed $n/2$ times. Let us call the variable counting the loops "i"; i goes from 1 to $n/2$.

When

$i = 1$: the first two subintervals limited by x_0, x_1, and x_2 are treated;

$i = 2$: the next two limited by x_2, x_3, and x_4 are treated;

...

$i = n/2$: the last two limited by x_{n-2}, x_{n-1}, and x_n are treated.

So, the contribution of two more subintervals is added to the integral of the function over $[a, b]$ during each loop. Using the variable G for that integral, a possible way to program this loop (with a Pascal-like code) is shown below:

```
G:= 0
h:= (b - a) / n
for i:= 1 to n/2  do
begin
        xlow :=  a + h . (2i - 2)
        xmid:=  a + h . (2i - 1)
        xup:= a + 2. h. i
        flow:= f(xlow)
        fmid:= f(xmid)
        fup:= f(xup)
        G:= G + h . (flow + 4 fmid + fup) / 3
end
```

It is assumed that the function f(x) has been programmed separately in another subprogram of the FUNCTION-type.

After this loop has been executed $n/2$ times, each couple of subintervals has added its contribution to the variable G, which contains at the end the total integral. This way of programming may be smoothed in several ways to reduce the number of calculations in each loop. The following piece of program computes the same integral a little faster:

```
G:= 0
h:= (b - a) / n
xlow:= a
flow:= f(a)
for i:= 1 to n/2 do
begin
        xmid:= xlow + h
        xup:= xmid + h
        fmid:= f(xmid)
        fup:= f(xup)
        G:= G + h (flow + 4 fmid + fup) / 3
        xlow:= xup
        flow:= fup
end
```

At the end of the loop, we have used the fact that the `xup` and `fup` values of a pair of subintervals are the same as the `xlow` and `flow` values for the next pair. The number of function calls is therefore reduced from 3 to 2 during each loop. This principle has been used in the chapter on individual stellar orbits in the galaxy (Chapter 12). We will present a detailed flowchart for that particular case then.

The following sample program is written in MicroSoft QuickBasic 4.5:

```
DECLARE FUNCTION f (x)
```

```
'
' Sample program to compute the integral of a function f(x) between
' a lower boundary a and an upper boundary b.
'
' The exact form of f(x) is placed in FUNCTION f(x) of this program
'

INPUT "lower boundary a : "; a
INPUT "upper boundary b : "; b
INPUT "number of subintervals used (EVEN NUMBER !) :"; n%

'compute number of pairs of subintervals and interval size
 m% = n% / 2
 h = (b - a) / n%

'initialize the cycle
 integral = 0
 xlow = a
 flow = f(xlow)

'during each loop of following cycle, the contribution of two more
'subintervals is taken into account

FOR i% = 1 TO m%

'    compute other meshpoints of two successive subintervals and their
'    function values

      xmid = xlow + h
      xup = xlow + 2 * h

      fmid = f(xmid)
      fup = f(xup)

'    compute and add contribution of these two subintervals
      integral = integral + h * (flow + 4 * fmid + fup) / 3

'    prepare for next loop
      xlow = xup
      flow = fup

NEXT i%

PRINT "Approximated integral : "; integral

END


FUNCTION f (x)
'
'  contains the function that has to be integrated
'

   f = EXP(-x) + 2 * x

END FUNCTION
```

Chapter 2

The Morphology of Comet Tails

2.1 Introduction

Comets are among the most spectacular celestial objects. Most of the ten or twenty comets that are discovered each year can be observed only with intermediate or large telescopes. Only a few comets are sufficiently bright for amateur telescopes. In this chapter we will present a model to reproduce the shape of a comet's tail during its passage near the Sun.

The orbits of comets are highly eccentric ellipses, parabolas, or even hyperbolas. The eccentricities of the elliptic orbits may be close to one, while most of the planets (except for Mercury and Pluto) stay below 0.10. Although the perihelia of comets are in general less than about one Astronomical Unit (A.U.), their aphelia may be situated at more than 100 A.U., far beyond the orbit of Pluto. The Sun is the dominant gravitational source of the solar system and also determines the motion of the comets. However, the orbit of a comet is sometimes considerably disturbed by the gravitational effects of large planets. Families of comets may be formed in this way. This effect will not be included in our model, as we are only concerned with the morphology of an individual comet in the neighborhood of the Sun. Every possible planetary effect is disregarded.

A comet in the neighborhood of the Sun consists of three parts:

1. **The nucleus** (sometimes called the head) is mostly observed as a bright point-like source. The diameter is rather small compared to the rest of the comet or to the size of a planet—from a few kilometers to a few hundreds of kilometers. Particles from the nucleus are heated as the comet approaches the Sun, evaporate or sublimate, and are ejected in the coma or the tail.

2. **The coma** is observed as a diffuse, more or less spherical cloud that

surrounds the nucleus. A typical diameter is 10,000 km to 100,000 km. Just like the tail, the coma exists only in the vicinity of the Sun. It is assumed that particles in the coma are not subject to any gravitational force of the nucleus.

3. **The tail** varies in both size and shape. Some comets never develop an observable tail, while others have a double tail with two clearly different components. For some other comets such as the well known comet Arend-Roland, discovered at the Royal Observatory of Belgium, an anti-tail was seen when the earth passed through the orbital plane of the comet. This was probably an illusion provoked by particles spread out in the orbital plane along the comet orbit. The tail of a comet normally has a length of a few tenths of an A.U. Some of the tails are curved, while others are straight. The earth has crossed the tail of a number of comets without any effect. Tails of comets may be classified under two groups:

   (a) *Type 1 tails* consist of ionized molecules of hydrogen, oxygen, nitrogen and carbon. They are very linear, with a detailed structure, pointing in the opposite direction of the Sun.

   (b) *Type 2 tails* show no detailed structure. They are dusty, containing small particles about the size of half a micron. They are more bent than Type 1 tails. A third type, Type 3, is sometimes referred to as an extreme subgroup of Type 2 tails.

The study of the morphology, the size, the orientation, the curvature and the evolution of the tail is important, not only for our knowledge of comet physics, but also for understanding the characteristics of the interplanetary medium and the solar wind. The particles in the tail are subject to a gravitational attraction of the Sun, but it is also assumed that a second force, representing the effects of the solar wind, is acting on the particles. It is assumed that this repulsive force is proportional to the inverse square of the distance to the Sun. Since the gravitation has the same dependency on the distance, it is possible to express the repulsive force in units of the gravitational attraction. A parameter μ, which is defined by the fact that the magnitude of the repulsive force is written as $1 - \mu$ times the gravitational force, is therefore introduced. Since the two forces can be added to each other, it is possible to consider the whole problem as if there were only gravitation powered by the Sun, but in that case gravitation is multiplied by a factor μ. By selecting various values of μ, it is possible to balance the relative importance of gravitation versus the effects of the solar interaction.

The equations of motion of particles in the tail, expressed in the two-dimensional space of the orbital plane, are then simply

$$x'' = -\mu \frac{x}{r^3} \tag{1}$$

and

$$y'' = -\mu \frac{y}{r^3} \tag{2}$$

in which r is the distance to the Sun, and (x, y) and (x'', y'') the position and acceleration in a Cartesian XY coordinate system in the orbital plane of the comet and origin in the Sun. The third power in the denominator may look strange but is compensated by the presence of a factor x or y in the numerator. In this way, the global dependency is the inverse square of the distance r, as it should be. In these formulae, which will not figure explicitly in the program, the unit of time is 58.132 days (one year divided by 2π); the unit of mass is the mass of the Sun; and the unit of length the Astronomical Unit. In this way the gravitational constant G becomes 1. The parameter μ is included to simulate the effect of repulsion and to make it possible to distinguish Type 1 from Type 2 tails. The parameter μ takes different values for each type.

It is not clear what happens with the particles once they have entered the tail, but there is no reason to assume that they will ever return to the nucleus. The particles of the tail are probably scattered along the orbit. In this way they may possibly be observed as an anti-tail when the earth crosses the orbital plane of a comet. The formation of a tail also means a loss of mass each time the comet passes the Sun. This attrition of mass may be a reason that the period of some comets slowly changes and that periodic comets don't reappear at the right moment.

2.2 The Orbital Elements

The formulae presented in this section are valid for every elliptic orbit. We'll apply them to comets but they are the same for planets. An ellipse is characterized by its eccentricity e and its semi-major axis a. When dealing with comets, it is also possible to work with the perihelion distance a_p instead of the semi-major axis. The perihelion distance is the minimum distance between the comet and the Sun during its passage. The relation between the three orbital parameters is

$$a_p = a(1 - e) \tag{3}$$

which follows from the definition of eccentricity. The eccentricity of an ellipse is always between zero and one. When it is zero, the ellipse reduces to a circle. Values near to one correspond to very eccentric ellipses. The ellipse becomes a parabola when the eccentricity becomes one. Hyperbolas have eccentricities larger than one. The so called orbital parameter p is given by

$$p = a(1 - e^2) = a_p(1 + e). \tag{4}$$

When a is given in A.U., p and a_p are also. The orbital period P expressed in years is then obtained from Kepler's law

$$P = a^{1.5} \tag{5}$$

One may always compute a and P, starting from the initial parameters e and a_p. The equation of an ellipse in polar coordinates with the origin in the Sun, and the X-axis pointing through the perihelion of the orbit of the comet is

$$r = \frac{p}{1 + e\cos\nu}. \tag{6}$$

In this expression ν is the true anomaly. Starting in the perihelion, it is measured in the direction of the motion of the comet. The distance to the Sun is r. When the comet is in its perihelion, the true anomaly equals zero, and r equals a_p. To simplify the calculations and the plotting of the results, it is better to rewrite (6) in normal rectangular coordinates. The X-axis as well as the origin remain the same, and the Y-axis points in the direction of $\nu = 90°$. The relations between polar and Cartesian coordinates are

$$x = r\cos\nu \tag{7}$$

and

$$y = r\sin\nu. \tag{8}$$

Once e and a_p are selected, p may be computed, and r, x, and y are obtained for any choice of the true anomaly by means of (6,7,8). Since a_p is expressed in A.U., the same holds for r, x, and y. The orbit of the comet near the Sun is drawn in the XY-plane for a number of ν-values.

2.3 Cometocentric Coordinates

We are able to fix the position of the nucleus of the comet for a given true anomaly with the formulae of section 2.2. The calculation of the shape of the tail is performed in a new coordinate system. The origin of this system moves with the nucleus along the orbit of the comet. We shall call this the ST-system. The S-axis points from the Sun along the radius vector through the nucleus of the comet, and the T-axis is perpendicular to S, but in the opposite direction of the motion of the comet. In this way the comet moves in the direction of negative T-coordinates. Figure 2.1 shows the two coordinate systems.

The position of a tail particle is computed in ST-coordinates and then converted to the fixed XY-system. Since the origin of the ST-system is moving with the comet, the conversion between the two coordinate systems will contain information on the position of the nucleus and will therefore depend on (x, y) also. The relations between the position of a tail particle in ST and XY coordinates are presented in (9,10). The position of the nucleus in XY is (x, y). A tail particle has coordinates (x', y') in XY and (s, t) in ST. The nucleus has trivial coordinates (0,0) in the ST-coordinate system.

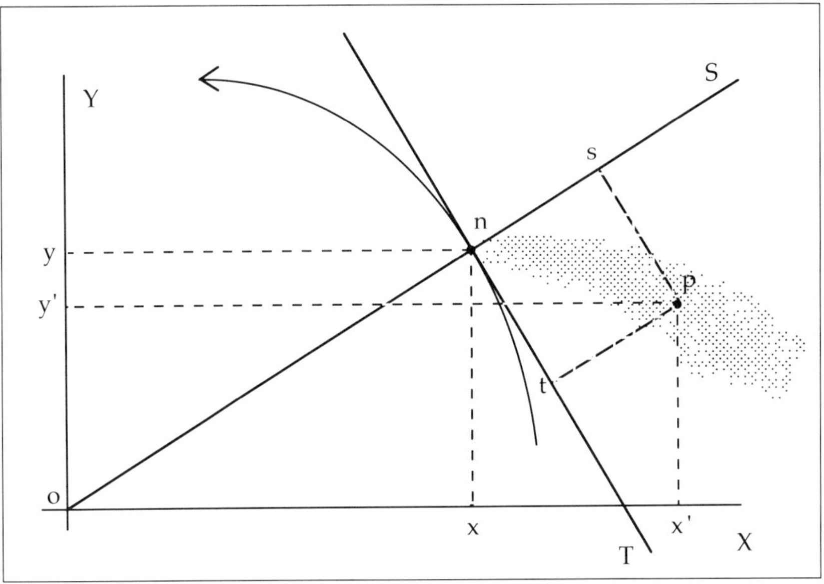

Fig. 2.1: *o is the position of the Sun which is the origin of the XY-system. n is the position of the nucleus having coordinates (x, y) in XY and (0,0) in ST. p is the position of a tail particle, with coordinates (x', y') in XY and (s, t) in ST.*

The ST-coordinate system is often presented with the T-axis tangential to the orbit of the comet. This is incorrect. One can easily show that the transformation formulae (9,10) or (11) convert an orthogonal coordinate system XY into an orthogonal system ST and vice-versa.

$$s = \frac{xx' + yy' - r^2}{r} \tag{9a}$$

$$t = \frac{yx' - xy'}{r} \tag{9b}$$

where

$$r = \sqrt{x^2 + y^2}. \tag{10}$$

The inverse relations are

$$x' = \frac{sx + ty + rx}{r} \tag{11a}$$

and

$$y' = \frac{sy - tx + ry}{r}. \tag{11b}$$

2.4 The Syndynames and the Bessel-Bredechin Theory

The origin of this theory goes back to the nineteenth century, with Bessel. It was later improved by many others such as the Russian astronomer Bredechin. It is assumed in this fundamental theory that the particles leaving the nucleus and entering the coma and the tail escape from the nucleus with a certain velocity g making an angle G with the S-axis. Positive values of G may be taken in the direction of the motion of the comet.

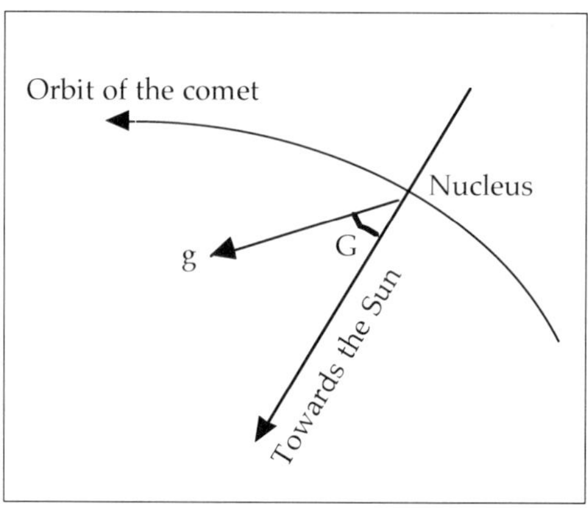

Fig. 2.2: *The angle G under which particles escape from the nucleus with velocity g.*

The coordinates (s, t) of a particle in the tail are then expressed as functions of the time. The time coordinate, which figures explicitly in the equations, is eliminated from these expressions, resulting in a new relation between s and t. One obtains in this way the equation of the position of all the particles which have left the core in the past with a velocity g under an angle G. Such a position may be considered as a streamline in the tail and is called a syndyname.

It is important to realize that a syndyname is not the path followed by one individual particle, but an intersection of a number of individual paths at a certain time, selecting at every path the position where the particle is at that moment. It is really a picture, at one particular moment, of all the particles that have left the core under equal circumstances. It is also possible to compute individual paths. Such a path is called a *synchrone*. It may be used to visualize explosive ejections of matter at one certain moment. They are observed as jetstreams in the tail near the coma.

At this point the mysterious force parameter μ re-enters the theory. For a comet with given force parameter, orbital elements and position in the orbit fixed by the value of ν, the equation of the syndyname with outflow velocity g under an angle G is

$$t = g\sin G\left(\frac{\sqrt{2}r}{\sqrt{1-\mu}}s^{1/2} - \frac{4er\sin\nu}{3(1-\mu)\sqrt{p}}s\right) + \frac{2\sqrt{2p}}{3r\sqrt{1-\mu}}s^{3/2}. \tag{12}$$

This equation is obtained after eliminating the time in the equations of motion of a particle leaving the nucleus. This deduction, however, is beyond the scope of this monograph. The computations to be performed are made more efficient by introducing the following intermediate steps:

$$A_1 = \frac{\sqrt{2}r}{\sqrt{1-\mu}} \tag{13a}$$

$$A_2 = \frac{4er\sin\nu}{3(1-\mu)\sqrt{p}} \tag{13b}$$

$$A_3 = \frac{2\sqrt{2p}}{3r\sqrt{1-\mu}}. \tag{13c}$$

Equation (12) then becomes

$$t = g\sin G(A_1 s^{1/2} - A_2 s) + A_3 s^{3/2}. \tag{14}$$

This modification is performed to reduce the number of calculations. The quantities A_1, A_2, and A_3 are only dependent on the basic parameters e, p, and μ and on the position (r, ν). Therefore, once the position of the nucleus is fixed, these three expressions may be computed and used to determine all the syndynames one wants to compute for a given point in the orbit. This shorter way of writing the equation of a syndyname also gives us some information of the structure of a syndyname.

The factor A_3 is independent of the outflow parameters g and G, and describes the position of the axis of the tail. It is in fact a syndyname with the outflow angle G equal to zero or 180°. The two other factors of the equation, containing A_1 and A_2 depend on g and G. They describe how much a syndyname deviates from the axis of the tail. Since G is only present as its sinus, the maximal deviations are found for $G = 90°$ and $-90°$. The factor g determines the width of the tail, since it acts as a multiplication factor on the deviation from the tail axis. It is therefore sufficient to compute the syndynames of 90° and $-90°$ to determine the boundary of the tail. Taking $G = 0$ gives us the axis. It is assumed that the outflow velocity g is constant during the whole computation of a tail and even during the whole passage near the Sun.

The table below lists typical generally-used values for g in combination with μ. In this table g is expressed in A.U./58.132 days so that the values listed can be used immediately in the program. The speed in kilometers/second may be obtained from g (in kilometers/second) $= 29.785\ g$.

	Type 1	Type 2	Type 3
g	0.20	0.05	0.02
$1 - \mu$	> 10	4–2.5	0.5–0.1

The values of g may be changed over a factor 2 in both directions.

2.5 Computational Method

The computation of the structure of the tail at various positions in the orbit is performed in four stages. Each of these stages is subordinate to the previous one. No iterative procedure such as the solution of a differential equation is needed. All calculations are straightforward. The four stages are as follows:

1. Select a comet. This means the choice of the eccentricity e, the perihelion distance a_p, and the repulsion parameter μ. With these choices the orbit and the effects of gravitation and repulsion are settled. The outflow velocity g may now also be selected in connection with the value of the force parameter by means of the table above.
2. Select a position in the orbit. This is performed by selecting a value of the true anomaly ν. We refer to the applications in section 2.6 for reasonable choices. It is now possible to compute the distance r to the Sun, and the rectangular coordinates of the nucleus (x, y). Furthermore the quantities A_1, A_2, and A_3 are now calculated with (13).
3. Select a syndyname. This is accomplished by making a choice for G, since each value of G determines a syndyname. It is an equation in the ST-system given in (14).

4. Compute for a number of s-values the corresponding t-coordinate of the syndyname. These (s, t) points are then converted into the heliocentric Cartesian coordinates (x', y') with (11). Plot all these (x', y') coordinates together with the position of the nucleus in the XY-coordinate system. One finds a position of the syndyname by connecting all the points.

As previously mentioned, each of these four stages is subordinate to the previous one. For one comet, various positions are considered. For every position, at least three syndynames are computed. Every syndyname is obtained by computing a number of (s, t)-points on it. For instance, taking nine positions, three syndynames at each position and six (s, t)-points at each of the syndynames, requires 162 points to be evaluated. These points, when plotted in the XY-axes, describe the shape of the tail and the nature of its changes when the comet is near the Sun. The choices for the syndyname are, of course, 0° or 180° for the tail axis, and 90° and −90° for the boundaries of the tail.

It is only meaningful to compute the length of the tail when the comet approaches the Sun, since at larger distances no tail is formed. Furthermore, the choice of s is also limited. Equation (14) makes it possible to compute a tail of any length, but in reality one should stop after a few times 0.1 A.U. Remember that the coordinates (s, t) are both expressed in A.U.

On programmable pocket calculators with label keys, the four stages may be written as independent programs using these keys. Computer programs in Basic or Pascal, for instance, may contain nested loops, every loop corresponding to one of the four stages of calculation:

Label	Input	Output
A	e, a_p, μ, and g	a, p, and P
B	ν	r, x, y, A_1, A_2, and A_3
C	G	
D	s	t, x', and y'

In this way, several positions may be computed by pressing B without having to compute each time the formula stored under A, etc. On larger calculators or a PC, For-loops such as in Pascal or equivalent instructions in Basic or Fortran may be used.

2.6 Applications

1. Tail of Type 2: Take

$$
\begin{aligned}
1 - \mu &= 1.00 \\
e &= 0.95 \\
a_p &= 0.50 \\
g &= 0.03
\end{aligned}
$$

The choice of the parameter shows that this is a dusty tail and will therefore have a considerable curvature.
Orbital parameters:

$$
\begin{aligned}
a &= 10 \text{ A.U.} \\
P &= 31.623 \text{ yrs.} \\
p &= 0.975 \text{ A.U.}
\end{aligned}
$$

Position of the comet: Take $\nu = -2$ radians $= -114.5916°$. Hence

$$
\begin{aligned}
r &= 1.612475 \\
x &= -0.6710264 \\
y &= -1.466219 \\
A_1 &= 2.280384 \\
A_2 &= -1.880871 \\
A_3 &= 0.577342
\end{aligned}
$$

We will first list the complete results for one syndyname. Consider the syndyname of $G = -90°$. Then

s	t	x'	y'
0.05	−0.01166371	−0.6812280	−1.5165381
0.10	−0.00901908	−0.7044401	−1.5609025
0.15	−0.00141906	−0.7321581	−1.6032046
0.20	0.00975923	−0.7631298	−1.6440177
0.25	0.02385544	−0.7967548	−1.6836164
0.30	0.04046857	−0.8326684	−1.7221678

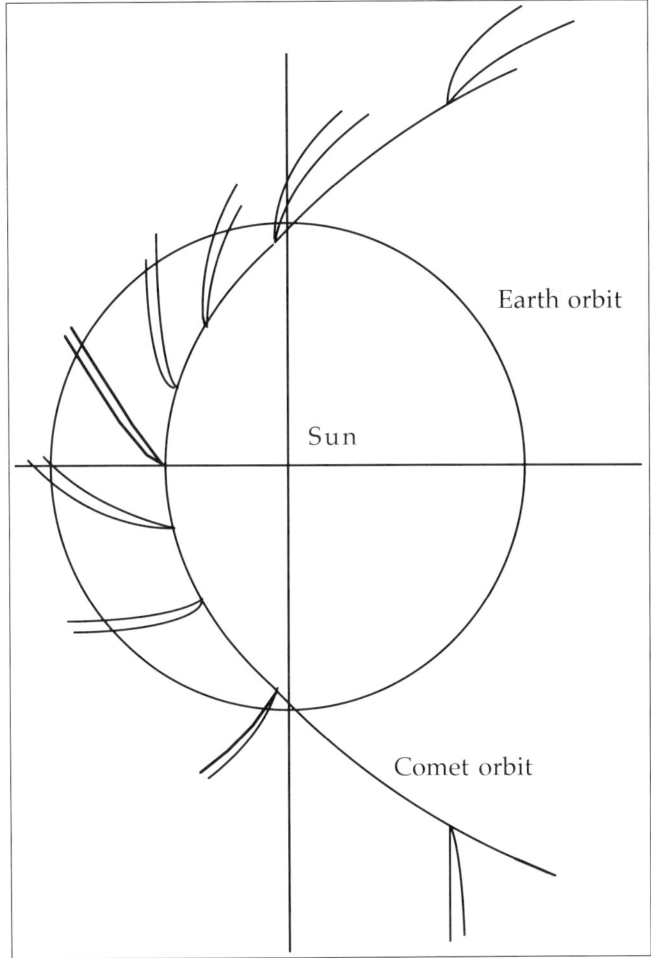

Fig. 2.3: *Tail of Type 2.*

This is only one syndyname at only one position. Figure 2.3 shows the complete results for three syndynames at nine positions of the nucleus ranging from $\nu = -2$ to $+2$ with steps of 0.5 radians.

2. Tail of Type 1: Take

$$1 - \mu = 18$$
$$e = 0.97$$
$$a_p = 0.50$$
$$g = 0.20$$

Hence

$$a = 16.7 \text{ A.U.}$$
$$P = 68.04 \text{ yrs.}$$
$$p = 0.985 \text{ A.U.}$$

Position of the comet: Take $\nu = 1$ radian $= 57.29577951°$. Hence

$$r = 0.6462859$$
$$x = 0.349189773$$
$$y = 0.5438308497$$
$$A_1 = 0.2154286$$
$$A_2 = 0.0393717$$
$$A_3 = 0.3412565$$

Consider the syndyname of $G = 90°$. Then

s	t	x'	y'
0.05	0.01305591	0.3871911	0.5788503
0.10	0.02362895	0.4231031	0.6152112
0.15	0.03533109	0.4599652	0.6509620
0.20	0.04821656	0.4978231	0.6860735
0.25	0.06223134	0.5366312	0.7205748
0.30	0.07731089	0.5763353	0.7545009

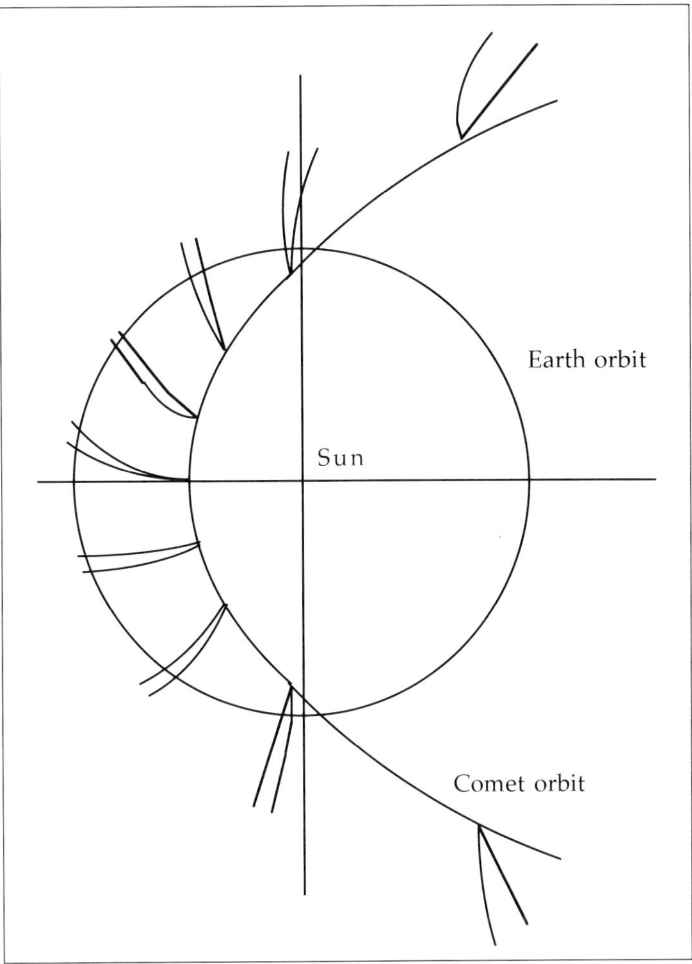

Fig. 2.4: *Tail of Type 1.*

Never try to fit your theoretical tails computed with this method to the tail of a comet shown on a picture—the picture will probably not be taken perpendicular to the orbital plane of the comet.

2.7 The Program Listing

The following sample program is written in MicroSoft QuickBasic 4.5:

```
CONST pi = 3.1415926536#
```

```
CLS
PRINT "Astrophysics With a PC : COMET TAILS"
PRINT "----------------------------------------"
PRINT ""
PRINT "------- Minimal solution program ---------"
PRINT " "
PRINT "Input parameters and orbital elements : "
INPUT "Perihelion distance (A.U.)      : ", ap
INPUT "Eccentricity of the comet orbit : ", ecc
INPUT "Parameter 1 - mu                : ", mu
INPUT "Outflow velocity                : ", g
p = ap * (1 + ecc)


FOR i% = -4 TO 4
'
'   this FOR-cycle considers 9 positions of the comet in its orbit
'
    CLS

'  next block computes position of the nucleus and parameters A1, A2 and A3
'  and shows these results on screen
    nu = .5 * i%
    r = p / (1 + ecc * COS(nu))
    x = r * COS(nu)
    y = r * SIN(nu)
    a1 = SQR(2 / mu) * r
    a2 = 4 * ecc * r * SIN(nu) / 3 / mu / SQR(p)
    a3 = SQR(8 * p / mu) / 3 / r

    PRINT USING "Position ##  : true anomaly = ##.###    r = ##.###_
   x = ##.###   y = ##.###"; i% + 5; nu; r; x; y
    PRINT USING "                a1 = ###.######        a2 = ###.######_
        a3 = ###.######"; a1; a2; a3
    PRINT " "
    PRINT "                              s              t              x_
            y"

    FOR j% = -1 TO 1
'
'       this FOR-cycle considers 3 syndynames for each of the 9 positions
'
        gg = j% * pi / 2
        ggdegree = j% * 90
        PRINT USING "Syndyname for G = ###.#"; ggdegree

        FOR k% = 1 TO 9
'
'           this FOR-cycle computes 9 points (s,t) and their (x',y')_
 transformation
'
            s = .05 * k%
            t = g * SIN(gg) * (a1 * SQR(s) - a2 * s) + a3 * s * SQR(s)
            xx = (s * x + t * y + r * x) / r
            yy = (s * y - t * x + r * y) / r
            IF k% MOD 2 = 1 THEN PRINT USING "                    ###.######_
     ###.######    ###.######    ###.######"; s; t; xx; yy
        NEXT k%
    NEXT j%

    LOCATE 23, 1
    PRINT "Press any key to continue "
    DO
    LOOP WHILE INKEY$ = ""
NEXT i%
END
```

Chapter 3

Meteor Dynamics

3.1 Introduction

The Earth, while proceeding in its large elliptic orbit around the Sun, continuously meets small particles drifting in the solar system. These particles normally follow Keplerian orbits in the gravitational field dominated by the Sun, but may be disturbed and attracted by a larger body such as a planet or a satellite. When attracted by the Earth, a particle may eventually enter the atmosphere. Its speed at that moment is very high—between 11 km/s and 72 km/s. The lower value is the escape velocity of the Earth. Any body falling towards the Earth, starting with a zero initial speed relative to the Earth, will be accelerated by the gravitational field of the Earth to that final speed of 11 km/s at the moment of contact with the atmosphere. The upper limit is the sum of the escape velocity of the Sun (42 km/s) in the vicinity of the Earth's orbit and the orbital velocity of the Earth (30 km/s). A particle falling towards the Sun from outside the solar system will be accelerated to 42 km/s when passing the Earth's orbit. If the path of that particle happens to be in the opposite direction of the Earth's speed at that moment, another 30 km/s is added to obtain the speed of the particle relative to the Earth.

When the meteoroid enters the atmosphere, it will be heated and start evaporating due to friction and collision with the atmosphere particles. This process will become more and more efficient as the meteoroid penetrates deeper and deeper in the atmosphere where the density increases. The heat released during this phenomenon is partially used to ionize atoms along the path of the meteoroid. When the electrons recombine with the ions, a small amount of energy is freed in the form of light. This is observed as a *meteor*, the well-known shooting star.

Eventually, meteoroids become large enough to partially survive their destructive flight through the atmosphere. Their remnants, called *meteorites*, may then be found on the Earth's surface. Meteorites are very important for our knowledge of the structure, the origin, and the evolution of the solar system.

The purpose of this chapter is to compute the behavior of a small meteoroid as it penetrates and interacts with the Earth's atmosphere. This approach is opposite to that of observers of meteors, who study the light curves and paths of the meteors in the sky in order to derive relevant properties of the meteoroid provoking the meteor. We start from a physical theory and try to reproduce light curves similar to the ones noted by meteor observers. The theory used here is limited to homogeneous, nonfragmentating meteoroids. Fragmentation means that the meteoroid breaks into pieces during the final seconds before entering the Earth's atmosphere. This may result in sudden changes in the luminosity or deviations from the normal path. The results will be obtained in the form of a table containing the horizontal coordinate x and the vertical coordinate y, the two corresponding velocity components u and v, the mass M of the meteoroid and the apparent magnitude m for an observer on the ground right below the meteoroid. We will have to solve a system of five coupled differential equations (for x, y, u, v, and M). This will be performed using the predictor corrector method of Heun described in the first chapter.

3.2 Physical Background

Due to the interaction of the meteoroid's surface with the particles of the atmosphere, the speed of the meteoroid will decrease. The meteoroid will suffer a loss of momentum (momentum = the product of its speed and its mass). The change of momentum is proportional to the amount of air displaced by the meteoroid in one second. This amount of air is the product of the volume of air occupied in one second by the meteoroid and the density of the atmosphere at that particular height. The volume itself is the product of the speed V of the meteoroid and its cross sectional area S, which is given by

$$S = AM^{2/3}\rho_m^{-2/3} \tag{1}$$

where

A is the shape coefficient, which must be included since we do not know the exact shape of the meteoroid. The shape coefficient for a perfect sphere is 1.21, and for bodies such as small ellipsoids or cones it is always near to one. A is a first free parameter.

M is the initial mass of the meteoroid. Its decrease in time will be described by a differential equation.

ρ_m is the density of the meteoroid, which is supposed to be homogeneous. The density may be a few grams/centimeter^3, but values near one are also possible, depending on the chemical composition of the meteoroid. This is a second free parameter.

The amount of air displaced by the meteoroid in one second is then

$$SV\rho_a \tag{2}$$

where

V is the speed of the meteoroid at that particular moment, and

ρ_a is the density of the atmosphere at that particular height. It is a decreasing function of the height y above the ground.

A suitable approximation is

$$\rho_a(y) = \exp\Big(-6.65125 - 13.9813\ 10^{-7}y\Big) \tag{3}$$

in which the height is expressed in centimeters and the density in grams/centimeter^3. The exponent 10^{-7} should be replaced by 10^{-2} if y is expressed in kilometers.

The loss of momentum is then expressed by the drag equation. This is a differential equation of the fourth order, since V has two components u and v, and the factor dV/dt is in fact the second derivative of the position. The drag equation in which we have inserted equation (1) is

$$dV/dt = -GAM^{-1/3}\Big(\rho_m^{-2/3}\Big)\Big(\rho_a V^2\Big) \tag{4}$$

where

$V^2 = u^2 + v^2$, the square of the velocity, and

$G =$ the drag coefficient. It is a proportional constant determining the efficiency of the resistance of the atmosphere.

Equation (4) has two components du/dt and dv/dt, the accelerations in the x- and y-direction. To make this clear, an XY-coordinate system is defined with the X-axis beneath the path of the meteoroid and the Y-axis pointing upwards through the initial position of the meteoroid at $t = 0$. The two components of (4) are

$$du/dt = -GAM^{-1/3}\Big(\rho_m^{-2/3}\Big)\rho_a\Big(u^2 + v^2\Big)^{1/2} u \tag{5}$$

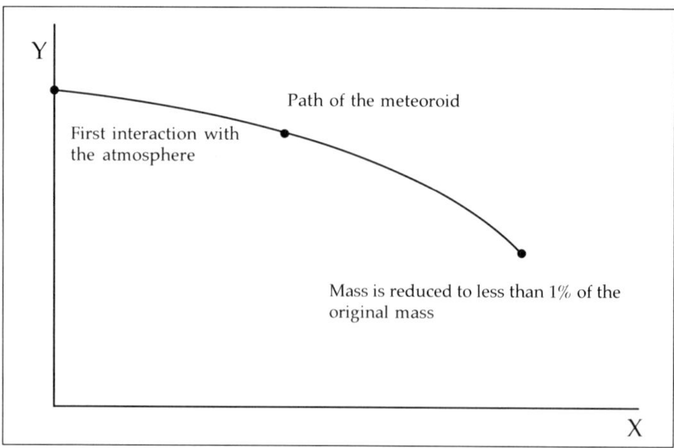

Fig. 3.1: *Coordinate system used in this chapter.*

and

$$dv/dt = -GAM^{-1/3}\left(\rho_m^{-2/3}\right)\rho_a\left(u^2+v^2\right)^{1/2} v - g. \tag{6}$$

The gravitational acceleration $g = 980\ \mathrm{cm/s^2}$ is included *pro forma*. Since the meteoroid survives only a few seconds in the atmosphere before it is totally burned up, the effect of gravitation on the vertical component of the speed is marginal. It is much smaller than the velocity decrease caused by friction and collision.

The change in the components of the position (the coordinates x and y) are then simply given by the equations

$$dx/dt = u \tag{7}$$

and

$$dy/dt = v. \tag{8}$$

So far we have given the four differential equations describing the trajectory of the meteoroid. Equations (5,6) for the change of the velocity components contain the mass M of the meteoroid. The variation of the mass as a function of time is given by a fifth differential equation. It is easily derived from the principle of conservation of energy:

$$dM/dt = -\frac{LA}{2\xi}M^{2/3}\left(\rho_m^{-2/3}\right)\rho_a\left(u^2+v^2\right)^{3/2} \tag{9}$$

where

ξ is the energy needed to evaporate one gram of the meteoroid, and

L is a parameter describing the efficiency of heat transfer.

Equations (5–9) are a system of five coupled differential equations that has to be solved numerically. The final calculation is the apparent magnitude m of the meteoroid as seen by an observer placed right below the meteoroid on the ground. The light that we see is in fact a part of the energy released by the evaporation of the meteoroid. That part is expressed by

$$E = -0.5\tau \frac{dM}{dt}\left(u^2 + v^2\right). \tag{10}$$

The parameter τ may be taken a few times 0.01.

Using the relation between the apparent magnitude, the absolute magnitude and the distance D (expressed in parsec) from the source of light to the observer

$$M = m + 5 - 5\log D$$

and comparing these quantities with the corresponding values for the sun, we see that the apparent magnitude of the meteor for an observer on the Earth's surface below the meteoroid is

$$m = 5\log y(\mathrm{cm}) - 2.5\log E - 8.795 \tag{11}$$

in which all logarithms are base 10. The coordinate y is expressed in centimeters, the normal unit of length of this chapter. With this formula, the observer moves over the Earth's surface with the same horizontal speed as the meteoroid. He will observe the meteor as fixed in his local zenith.

We need to select five initial conditions (initial mass, x-position, y-position, x-velocity, and y-velocity) and six free parameters (A, L, G, ξ, τ, and ρ_m) to obtain a complete description of a meteoroid. For practical use in the program, it is possible to work with three parameters since the six mentioned above appear only in certain combinations in the equations. We will therefore combine the free parameters to

$$K1 = \frac{GA}{\rho_m^{2/3}} \tag{12a}$$

and

$$K2 = \frac{LA}{2\xi\rho_m^{2/3}} \tag{12b}$$

with τ as the third parameter.

Typical values for $K1$ are between 0.5 and 1, and for $K2$ a few times 10^{-11}. Values for the individual free parameter may be found in professional textbooks on planetary astrophysics.

3.3 Summary

Let us first introduce shorter notations for the derivatives dx/dt, dy/dt, du/dt, dv/dt, and dM/dt, and replace them by x', y', u', v', and M'. In this way, and by using the constants defined by (12a,b), the five differential equations of motion become

$$\begin{aligned} x' &= u = f_x \\ y' &= v = f_y \\ u' &= -K1\rho_a V M^{-1/3}\, u = f_u \\ v' &= -K1\rho_a V M^{-1/3}\, v - g = f_v \\ M' &= -K2\rho_a V^3 M^{2/3} = f_M \end{aligned}$$

in which

$$V = \left(u^2 + v^2\right)^{1/2}$$

and ρ_a is given by (3).

For the apparent magnitude:

$$\begin{aligned} E &= -0.5\tau f_M V^2 \\ m &= (11). \end{aligned}$$

The five initial conditions at time $t = 0$ needed to start the numerical solution of the equations follow:

x_o = the initial horizontal coordinate which may always be taken to be zero. In this way the x-coordinate is measured starting from the x-position where the meteoroid has its first contact with the atmosphere.

y_o = the initial height above the Earth's surface. It should be taken high enough so that the total process of ablation may be followed. It is therefore necessary to start from heights between 120 km and 160 km, or in other words 1.2×10^7 cm to 1.6×10^7 cm. Initial values higher than 160 km are meaningless since the density of the atmosphere is too low above that height. For the opposite reason we should stay above 120 km.

u_o = the initial horizontal velocity component, expressed in centimeters/second.

v_o = the initial vertical velocity component. This component is negative since the meteoroid moves down while the Y-axis has been defined upwards. Since entry velocities are between 11 km/s and 72 km/s, one should take

$$1.1 \times 10^6 \text{cm/s} < \left(u^2 + v^2\right)^{1/2} < 7.2 \times 10^6 \text{cm/s}.$$

For practical use it is better to direct the motion of the meteoroid mainly downwards. This means that we should take v_o (in absolute value) larger than u_o, always taking into account the boundaries on the global velocity and the fact that v_o is negative.

M_o = the initial mass of the meteoroid at time $t = 0$. We may, for instance, consider values between 0.01 g and 1 g. It is possible to obtain meteoroids yielding a small remnant on the ground when starting from larger values.

We refer to section 3.2 for typical values for the free parameters. All these are given in centimeter-gram-second units so that they can be applied without any changes.

3.4 Numerical Method

The main purpose of the algorithm is to compute the new state x_{i+1}, y_{i+1}, u_{i+1}, v_{i+1}, and M_{i+1} at time t_{i+1}, starting from the values at t_i. We will solve this problem using Heun's predictor-corrector method, which was described in Chapter 1. This means that we have to evaluate the five functions f_x, f_y, f_u, f_v, and f_M twice each step, once in the predictor and once in the corrector. It is, therefore, better to place these five functions in a subprogram. The apparent magnitude is computed after the new situation at t_{i+1} has been determined. In the following description of the numerical procedure we will write the functions f_x, f_y, f_u, f_v, and f_M with (i) if they are to be evaluated at the state x_i, ..., M_i, with $(\circ_{i+1})$ for the predicted state $x\circ_{i+1}$, $y\circ_{i+1}$, $u\circ_{i+1}$, $v\circ_{i+1}$, and $M\circ_{i+1}$ and with $(i+1)$ for the corrected new state x_{i+1}, y_{i+1}, u_{i+1}, v_{i+1}, and M_{i+1}.

The computations to be performed are then

$$t_{i+1} = t_i + dt \tag{13}$$

$$\rho_{a,i} = \ldots \text{compute from (3) in } y_i$$

$$V_i = \left(u_i^2 + v_i^2\right)^{1/2}$$

and for the predictor

$$\begin{aligned}
f_{x,i} &= u_i \\
f_{y,i} &= v_i \\
f_{u,i} &= -K1\,\rho_{a,i}\,V_i\left(M_i^{-1/3}\right)u_i \\
f_{v,i} &= -K1\,\rho_{a,i}\,V_i\left(M_i^{-1/3}\right)v_i - 980 \\
f_{M,i} &= -K2\rho_{a,i}\left(V_i^3\right)\left(M_i^{2/3}\right).
\end{aligned}$$

Then compute

$$\begin{aligned}
x_{\circ i+1} &= x_i + dt\,f_{x,i} \\
y_{\circ i+1} &= y_i + dt\,f_{y,i} \\
u_{\circ i+1} &= u_i + dt\,f_{u,i} \\
v_{\circ i+1} &= v_i + dt\,f_{v,i} \\
M_{\circ i+1} &= M_i + dt\,f_{M,i}.
\end{aligned}$$

These five expressions give the predicted situation at time t_{i+1}.

Then

$$\begin{aligned}
\rho_{a,\circ i+1} &= \ldots \text{ compute from (3) in } y_{\circ i+1} \\
V_{\circ i+1} &= \left(u_{\circ i+1}^2 + v_{\circ i+1}^2\right)^{1/2} \\
f_{x,\circ i+1} &= u_{\circ i+1} \\
f_{y,\circ i+1} &= v_{\circ i+1} \\
f_{u,\circ i+1} &= -K1\,\rho_{a,\circ i+1}\,V_{\circ i+1}\left(M_{\circ i+1}\right)^{-1/3} u_{\circ i+1} \\
f_{v,\circ i+1} &= -K1\,\rho_{a,\circ i+1}\,V_{\circ i+1}\left(M_{\circ i+1}\right)^{-1/3} v_{\circ i+1} - 980 \\
f_{M,\circ i+1} &= -K2\,\rho_{a,\circ i+1}\left(V_{\circ i+1}\right)^3\left(M_{\circ i+1}\right)^{2/3}
\end{aligned}$$

and for the corrector

$$\begin{aligned}
x_{i+1} &= x_i + 0.5dt(f_{x,\circ i+1} + f_{x,i}) \\
y_{i+1} &= y_i + 0.5dt(f_{y,\circ i+1} + f_{y,i}) \\
u_{i+1} &= u_i + 0.5dt(f_{u,\circ i+1} + f_{u,i}) \\
v_{i+1} &= v_i + 0.5dt(f_{v,\circ i+1} + f_{v,i}) \\
M_{i+1} &= M_i + 0.5dt(f_{M,\circ i+1} + f_{M,i}) \\
E_{i+1} &= \ldots \text{with (10) using } f_{M\circ i+1} \text{ for } dM/dt \text{ and } u_{\circ i+1} \text{ and } v_{\circ i+1} \\
m_{i+1} &= \ldots \text{with (11) using } E_{i+1} \text{ and the data of state } i+1.
\end{aligned}$$

During the calculations the time step dt has to be decreased a few times. If we continue using the same time step of 0.1 second (a typical time step for the initial part of the flight in the atmosphere) we would encounter numerical difficulties during the second part of the flight when the characteristics of the meteoroid change very rapidly. With a time step of 0.1 we would miss the details during the last few tenths of a second of the trajectory. Starting from a small step of 0.01, on the other hand, would require too many steps during the first part. A possible solution is as follows:

$$\begin{aligned}
dt &= 0.1 \text{ sec. as long as the mass is larger than 80\% of the initial mass,} \\
&= 0.05 \text{ sec. while the mass is between 80\% and 50\%,} \\
&= 0.02 \text{ sec. while the mass is between 50\% and 35\%,} \\
&= 0.01 \text{ sec. as soon as the mass has dropped below 35\%.}
\end{aligned}$$

The computations stop as soon as the mass has dropped below 1% of the initial mass. All meteoroids considered here will produce a similar light curve. The magnitude observed from the Earth will always increase in a linear way, followed by a steep decrease in a fraction of a second. Other light curves cannot be obtained with our simple program since they are the result of nonhomogeneous and/or fragmentating meteoroids. Such model calculations would require additional physics and free parameters to describe the internal structure of the meteoroid.

It should be stressed that the apparent magnitude calculated with (9,10) is for an observer permanently located directly under the meteoroid, hence an observer travelling over the surface of the Earth with the same horizontal speed. Considering the great speed of the meteoroid, even in the horizontal direction, this seems to be an unrealistic situation. However, the difference between the light curve of the static observer and the one of the co-moving observer is rather small, since the motion and the ablation of the meteoroid take place at very large heights. To obtain the apparent magnitude curve for the static observer located directly under the initial position of the meteoroid, it is sufficient to replace the factor y in equation (11) by the square root of

$(x^2 + y^2)$. All our examples are computed with the magnitude formula given by (11).

3.5 Applications

As a first example and test for your program, here we present the complete results for a meteoroid with the following initial conditions and free parameters:

$$
\begin{aligned}
x_o &= 0 \\
y_o &= 160\,\text{km} = 1.6 \times 10^7\,\text{cm} \\
u_o &= 20\,\text{km/s} = 2.0 \times 10^6\,\text{cm/s} \\
v_o &= -40\,\text{km/s} = -4.0 \times 10^6\,\text{cm/s} \\
M_o &= 0.01\,\text{g} \\
K1 &= 1 \\
K2 &= 1.10^{-11} \\
\tau &= 0.02
\end{aligned}
$$

Although the internal units of the program are centimeters and grams, we have listed the position coordinates and velocity components in kilometers and kilometers/second. However, do not forget that the initial conditions should be given in centimeters and centimeters/second to be consistent with the typical values of the free parameters.

The meteoroid in this example is visible to the naked eye during 0.7 seconds (a magnitude brighter than +5), see Table 3-1. The magnitude reaches its maximum after about 1.5 seconds. After that, the meteoroid disappears in only 0.2 seconds. The speed of the meteoroid decreases only marginally, which is a consequence of the very short time that the meteor survives. For the same reason gravity has virtually no effect on the vertical velocity.

A static observer right below the point where the meteoroid enters the atmosphere will see it for the first time ($m > 5$) at a height of about 81° above the horizon. This height drops to 69° when the meteor vanishes 0.7 seconds later. These angles may be checked by computing the value of $\arctan(y/x)$.

We have computed with the same program some other meteoroids. Their initial conditions and free parameters are listed in Table 2. The meteoroid labelled "B" is the one for which complete results are given. All initial conditions are in centimeters, centimeters/second, and grams. The light curves for these cases are plotted as a function of the time. As mentioned above, all the curves are qualitatively the same.

Table 3-1

t	x	y	u	v	mass	m
0.10	2.0000	156.0000	19.99999	−40.00095	0.0099986	10.77
0.20	4.0000	151.9998	19.99996	−40.00188	0.0099961	10.11
0.30	6.0000	147.9996	19.99992	−40.00277	0.0099917	9.45
0.40	8.0000	143.9993	19.99984	−40.00360	0.0099841	8.78
0.50	10.0000	139.9989	19.99971	−40.00431	0.0099708	8.11
0.60	11.9999	135.9984	19.99947	−40.00483	0.0099476	7.44
0.70	13.9999	131.9979	19.99906	−40.00499	0.0099071	6.77
0.80	15.9997	127.9974	19.99835	−40.00454	0.0098364	6.10
0.90	17.9995	123.9970	19.99709	−40.00301	0.0097135	5.44
1.00	19.9992	119.9968	19.99488	−39.99956	0.0095010	4.77
1.10	21.9985	115.9971	19.99098	−39.99273	0.0091368	4.12
1.20	23.9973	111.9983	19.98403	−39.97982	0.0085224	3.48
1.30	25.9953	108.0011	19.97149	−39.95571	0.0075160	2.88
1.35	26.9937	106.0037	19.96189	−39.93698	0.0068285	2.62
1.40	27.9915	104.0074	19.94870	−39.91109	0.0059874	2.38
1.45	28.9885	102.0126	19.93035	−39.87487	0.0049890	2.17
1.47	29.3871	101.2153	19.92111	−39.85657	0.0045504	2.11
1.49	29.7854	100.4183	19.91044	−39.83542	0.0040922	2.05
1.51	30.1835	99.6219	19.89806	−39.81085	0.0036186	2.00
1.53	30.5813	98.8259	19.88360	−39.78213	0.0031354	1.97
1.54	30.7801	98.4282	19.87548	−39.76596	0.0028921	1.97
1.55	30.9788	98.0306	19.86665	−39.74839	0.0026494	1.96
1.56	31.1774	97.6332	19.85703	−39.72925	0.0024083	1.97
1.57	31.3760	97.2360	19.84653	−39.70834	0.0021702	1.97
1.58	31.5744	96.8390	19.83503	−39.68543	0.0019366	1.99
1.59	31.7727	96.4423	19.82239	−39.66024	0.0017090	2.02
1.60	31.9708	96.0459	19.80846	−39.63245	0.0014891	2.05
1.61	32.1688	95.6497	19.79301	−39.60164	0.0012786	2.10
1.62	32.3667	95.2538	19.77580	−39.56731	0.0010793	2.16
1.63	32.5643	94.8583	19.75652	−39.52883	0.0008931	2.23
1.64	32.7618	94.4632	19.73476	−39.48539	0.0007216	2.33
1.65	32.9590	94.0686	19.70998	−39.43590	0.0005666	2.44
1.66	33.1560	93.6745	19.68145	−39.37893	0.0004296	2.59
1.67	33.3527	93.2810	19.64817	−39.31244	0.0003118	2.77
1.68	33.5490	92.8883	19.60865	−39.23346	0.0002140	3.00
1.69	33.7448	92.4963	19.56058	−39.13738	0.0001367	3.29
1.70	33.9402	92.1055	19.50003	−39.01633	0.0000793	3.69

Table 3-2

Case	x_o	y_o	u_o	v_o	M_o	$K1$	$K2$	τ
A	0	1.6×10^7	2×10^6	-4.10^6	1.00	1.00	1×10^{-11}	0.02
B	0	1.6×10^7	2×10^6	-4.10^6	0.01	1.00	1×10^{-11}	0.02
C	0	1.6×10^7	2×10^6	-6.10^6	1.00	1.00	1×10^{-11}	0.02
D	0	1.6×10^7	2×10^6	-4.10^6	1.00	1.00	5×10^{-11}	0.02
E	0	1.4×10^7	1×10^6	-3.10^6	0.01	0.80	1×10^{-11}	0.01

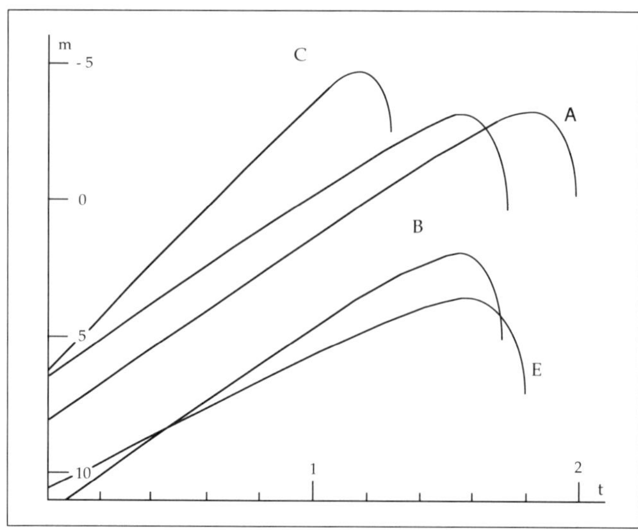

Fig. 3.2: *Apparent magnitude curves for the five meteoroids A to E.*

Repeat the computations of meteoroid B, starting from the same parameter values and initial conditions except for the initial mass. Increase that initial mass until a meteoroid that hits the ground is obtained.

3.6 The Program Listing

The following sample program is written in MicroSoft QuickBasic 4.5:

```
DECLARE FUNCTION datm (y)
DECLARE SUB effes (x, y, u, v, m, k1, k2, tau, fx, fy, fu, fv, fm, s)

'  This program is a minimal solution version of METEOR.PAS

CLS
PRINT "Astrophysics With a PC : METEOR"
PRINT "-----------------------------------"
PRINT ""
PRINT "-------- Minimal solution program --------"
PRINT ""
PRINT "Input of initial conditions and parameters : "
INPUT "Initial height (km)                : ", y
INPUT "Initial horizontal speed (km/s) : ", u
INPUT "Initial vertical speed (km/s)   : ", v
INPUT "Initial mass (gram)                : ", m
INPUT "Parameter K1 : ", k1
INPUT "Parameter K2  : ", k2
INPUT "Parameter tau : ", tau


' transform input data x,y,u and v from km to cm
' and make sure that v is negative
y = y * 100000!
```

```
x = 0
u = u * 100000!
v = -ABS(v * 100000!)
minit = m
t = 0!
i% = 1
PRINT ""
PRINT "  i   t       x          y          u          v          m        mag"
PRINT " "


DO  'This is the main loop of the program

' next 3 nested IF-blocks select the time step dt
    IF m > .8 * minit
    THEN
        dt = .1
    ELSE
        IF m > .5 * minit
        THEN
            dt = .05
        ELSE
            IF m > .35 * minit
            THEN
                dt = .02
            ELSE
                dt = .01
            END IF
        END IF
    END IF

    t = t + dt


' next line computes right hand sides of the 5 differential equations_
 (state i)
    CALL effes(x, y, u, v, m, k1, k2, tau, fx, fy, fu, fv, fm, s)

    ' next 5 instructions compute the predicted state ˚i+1
    x1 = x + dt * fx
    y1 = y + dt * fy
    u1 = u + dt * fu
    v1 = v + dt * fv
    m1 = m + dt * fm
    rho1 = datm(y1)
    s1 = SQR(u1 * u1 + v1 * v1)


' next line computes right hand sides of the 5 differential equations_
    (state ˚i+1)
    CALL effes(x1, y1, u1, v1, m1, k1, k2, tau, fx1, fy1, fu1, fv1, fm1, s1)


' next 5 instructions compute the corrected state at ˚i+1
    x = x + .5 * dt * (fx + fx1)
    y = y + .5 * dt * (fy + fy1)
    u = u + .5 * dt * (fu + fu1)
    v = v + .5 * dt * (fv + fv1)
    m = m + .5 * dt * (fm + fm1)


' Computation of the apparent magnitude :
    e = -.5 * tau * fm1 * s1 ^ 2
    mag = 5 * LOG(y) / LOG(10) - 2.5 * LOG(e) / LOG(10) - 8.795


' Results are shown on screen :
    IF y > 0 THEN
        PRINT USING "### ##.## ####.#### #####.#### #####.#### #####.####_
```

```
 ##.####### ####.##"; i%; t; x / 100000!; y / 100000!; u / 100000!;_
 v / 100000!; m; mag
    ELSE
        PRINT ""
        PRINT "Meteoroid has reached the ground"
        m = 0
    END IF
    i% = i% + 1


    IF i% MOD 15 = 0 THEN
' this IF makes the program pause after every 15 iterations
        PRINT ""
        PRINT "press any key to continue"
        DO
        LOOP WHILE INKEY$ = ""
        PRINT ""
    END IF

LOOP UNTIL m < minit * .01             'End of main loop

PRINT ""
PRINT "Program terminated. Press any key"
DO
LOOP WHILE INKEY$ = ""
END


FUNCTION datm (y)
'
'    computes the atmospheric density at height y(cm)
'
     datm = EXP(-6.65125 - 1.39813E-06 * y)
END FUNCTION


SUB effes (x, y, u, v, m, k1, k2, tau, fx, fy, fu, fv, fm, s)
'
'    computes fx, fy, fu, fv, fm and s (the speed) for a position (x,y)
'    velocity (u,v) and mass (m), and with parameters k1,k2 and tau
'
     fx = u
     fy = v
     s = SQR(u ^ 2 + v ^ 2)
     rho = datm(y)
     fu = -k1 * rho * s * u * EXP(-1 / 3 * LOG(m))
     fv = -k1 * rho * s * v * EXP(-1 / 3 * LOG(m)) - 980
     fm = -k2 * rho * s ^ 3 * EXP(2 / 3 * LOG(m))
END SUB
```

Chapter 4

The Restricted Three-Body Problem

4.1 Introduction

The general three-body problem is concerned with determining the resulting orbit for each of three bodies with given mass, initial position, and initial velocity vector moving in a three-dimensional space under the effect of the gravitational field created by their own masses.

It is not possible to solve this problem analytically because of insurmountable mathematical complications. Indeed, this problem is described by a system of differential equations of the eighteenth-order, namely nine second-order equations (three bodies move in three dimensions, and for each direction a second derivative is needed for the acceleration of that particular body in that particular direction).

Even after some important mathematical simplifications, accomplished by taking into account some physical laws of invariance and by selecting a "suitable" coordinate system, the problem remains too difficult for the capabilities of modern analysis.

Fortunately, a very special and much simpler case of the general problem has more applications in astronomy than the general problem itself. The simplifications involved are both mathematical and physical. This special case, called *the restricted three-body problem,* is the subject of this chapter. It is defined as follows: Two bodies (called the primaries) move in circular orbits around their common center of gravity. A third body with an infinitesimally small mass moves in the orbital plane of the two primaries under the effect of their gravitational field. The third body itself has no gravitational effect on the orbits of the primaries, due to its mass being virtually zero. We will briefly discuss the effect of the three restrictions of this problem.

1. The mass of the third body is virtually zero, making it possible to assume that the third body has no gravitational effect on the orbits of

the primaries. Therefore the orbits of the primaries may be determined as if the third body were absent, with the consequence that they obey the laws of the general two-body problem (meaning that their orbits are Keplerian ellipses in a fixed orbital plane).

2. Circular orbits for the primaries may be treated in a much simpler way than ellipses because they have a constant radius and a constant angular and orbital velocity. In an elliptical orbit, the orbital velocity becomes larger when the two bodies move closer to each other and decreases again when they move away.

3. The third body moves in the orbital plane of the primaries. This restriction implies an important mathematical simplification, since now the entire description of the problem becomes two-dimensional.

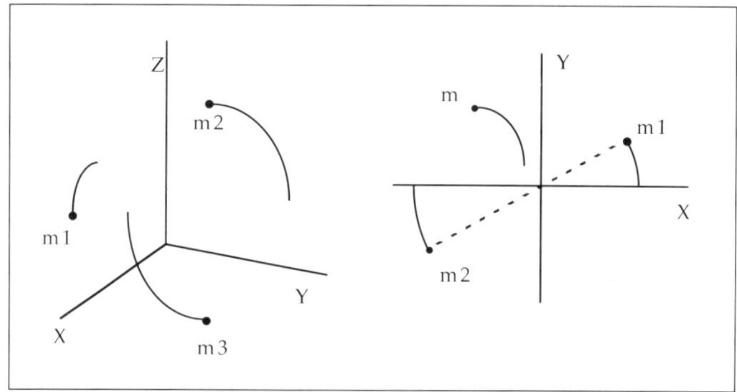

Fig. 4.1: *The general problem (left) and the restricted problem (right).*

Various textbooks have been published on the restricted three-body problem. *Theory of Orbits* by V. Szebehely is an excellent textbook on this subject. The interested reader will there find a detailed mathematical and physical description.

4.2 Some Astrophysical Applications

Any application should obviously be as consistent as possible with the three special characteristics of the restricted problem. It is, of course, impossible to find real applications that satisfy these severe demands exactly. The third body will always affect the other bodies and in this way disturb their elliptical motion. Circular orbits never occur, although eccentricities may be very close to zero. This is certainly the case for most of the orbits in the solar system except, of course, for comets. Finally, the third restriction on the coplanarity is almost never fulfilled, although the deviations from the main orbital plane are often very small. Gravitational systems with a dominating central mass show a tendency to evolve in a coplanar state as illustrated by the planets moving around the Sun and the satellite systems around the large planets. A number of examples which nearly meet the three restrictions may be found in astronomy. The theory of the restricted three-body problem may then be applied as a first approximation to obtain some general idea on the orbit of the third body. In some particular applications, this first approximation describes reality very well.

1. **Sun, Earth, and Moon.** One can ignore the eccentricity of the orbit of the Earth around the Sun, but the mass of the Moon, only about 80 times less than the Earth, can certainly not be considered zero. In fact, the Earth and the Moon orbit around a common center of gravity, which lies just inside the surface of the Earth. The inclination of the orbital plane of the Moon on the orbital plane of the Earth has to be included when detailed results are needed.

2. **Earth, Moon, and artificial satellite.** First, the satellite should be placed in the orbital plane of the Moon. The condition on the mass of the third body is certainly satisfied, and the eccentricity of the lunar orbit (only 0.0549) may be ignored in a first approximation for orbits around the Earth. In reality, the orbit of the satellite is disturbed by the Sun and the major planets and by nonspherical effects in the gravitational field of the Earth and the Moon.

3. **Sun, Jupiter, and asteroid.** Here, also, the coplanarity could be a problem and the perturbation by other planets (especially Saturn) may be important. An excellent example is the Trojan asteroids. Their orbits will be computed at the end of this chapter to show that the restricted problem may be used to describe actual situations with very reasonable accuracy.

4.3 Mathematical Description

We will start in a very fundamental way with Newton's law of gravitation. The gravitational force between two bodies with masses M_a and M_b on a distance R from each other is

$$F = G\frac{M_a M_b}{R^2} \tag{1}$$

in which G is the gravitational constant

$$G = 6.673 \times 10^{-8}\ \mathrm{cm}^3/\mathrm{g}/\mathrm{s}^2.$$

This formula is very important since all the other formulae depend upon it. The following notations will be used:

m is the mass of the third body,

m_1 is the mass of the first primary (the more massive one),

m_2 is the mass of the second primary (the less massive one),

(x, y) are coordinates of the third body,

(x_1, y_1) are coordinates of the first primary, and

(x_2, y_2) are coordinates of the second primary.

The origin of the coordinate system (XY) lies in the center of gravity of the two primaries, or in other words, in the center of the circular orbits of the primaries. Therefore, the orbital equations of the two primaries are

$$x_1 = a\cos(\omega\, t), \tag{2a}$$

$$y_1 = a\sin(\omega\, t), \tag{2b}$$

$$x_2 = -b\cos(\omega\, t), \tag{2c}$$

and

$$y_2 = -b\sin(\omega\, t). \tag{2d}$$

In these equations, a and b are the orbital radii of the primaries and ω is their angular velocity. Since the origin of the coordinate system is the center of gravity we also have the relation

$$m_1 a = m_2 b. \tag{3}$$

We will call the distances between the third body and the two primaries R_1 and R_2. It is obvious that R_1 and R_2 will be variable in time as the third body moves along its orbit around and between the primaries.

The expressions for the squares of R_1 and R_2 are

$$R_1 = \sqrt{(x - x_1)^2 + (y - y_1)^2} = \sqrt{(x - a\cos\omega\, t)^2 + (y - a\sin\omega\, t)^2} \tag{4}$$

and

$$R_2 = \sqrt{(x - x_2)^2 + (y - y_2)^2} = \sqrt{(x + b\cos\omega\, t)^2 + (y + b\sin\omega\, t)^2}. \tag{5}$$

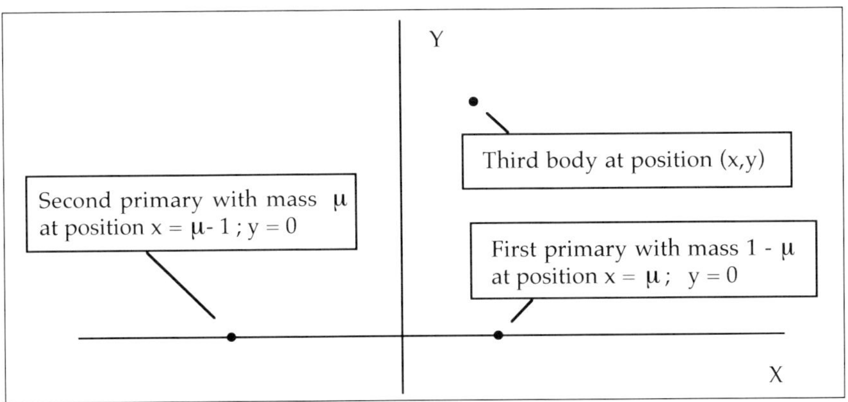

Fig. 4.2: *The orbital plane and coordinate system.*

We will not pay too much attention to the detailed mathematical deduction of the formulae; rather, we present only the most important intermediate steps. Since the coordinates of the third body are (x, y), we will use (x', y') for its velocity components and (x'', y'') for the acceleration components. Newton's law then gives the equations of motion of the third body:

$$x'' = -Gm_1 \frac{x - a\cos\omega\, t}{R_1^3} - Gm_2 \frac{x + b\cos\omega\, t}{R_2^3} \tag{6}$$

and

$$y'' = -Gm_1 \frac{y - a\sin\omega\, t}{R_1^3} - Gm_2 \frac{y + b\sin\omega\, t}{R_2^3}. \tag{7}$$

These equations are simplified in two ways without creating new restrictions. First, in most applications, one is only interested in the relative orbit of the third body compared to the positions of the primaries. Therefore, we may give the XY-coordinate system the same angular velocity as the primaries so that it co-rotates with them. In this co-rotating coordinate system, the primaries now have fixed positions. The origin of this new coordinate system is still the center of gravity, and the X-axis is the line connecting the two primaries. Its positive part is in the direction of the most massive primary. The Y-axis is perpendicular on the X-axis and crosses it in the center of gravity. In the resulting formulae, the trigonometric expressions for the motion of the primaries have vanished, but some new factors containing the angular velocity components x' and y' appear due to the pseudo forces induced by the rotating coordinate system. The appearance of these factors is purely due to the use of the new coordinate system and has nothing to do with the three-body problem itself. The factors appear in every rotating system. The new equations of motion in rotating coordinates (x, y) are

$$x'' = -Gm_1 \frac{x-a}{R_1^3} - Gm_2 \frac{x+b}{R_2^3} + \omega^2 x + 2\omega y' \tag{8}$$

and

$$y'' = -Gm_1 \frac{y}{R_1^3} - Gm_2 \frac{y}{R_2^3} + \omega^2 y - 2\omega x' \tag{9}$$

with

$$R_1 = \sqrt{(x-a)^2 + y^2} \tag{10}$$

and

$$R_2 = \sqrt{(x+b)^2 + y^2}. \tag{11}$$

The coordinates of the first primary are fixed at $(a, 0)$; for the second, they are fixed at $(-b, 0)$.

A second simplification enables us to use one set of equations of motion (and therefore one program) whatever the masses of the primaries are. The resulting formulae are applicable whether they are used to work in a Sun-Earth-Moon system, or an Earth-Moon-satellite system, or any other case. Instead of the normal units of length, mass, and time, new units are introduced. The new unit of length is the distance between the primaries. This change is possible since the circular orbits guarantee that this distance is constant. A similar situation is found in the solar system where the average distance between the Sun and the Earth, the astronomical unit, is used as a unit of length. The new unit of mass is the sum of the masses of the primaries. The reduced masses of the primaries then become

$$\mu_1 = \frac{m_1}{m_1 + m_2}$$

and

$$\mu_2 = \frac{m_2}{m_1 + m_2}.$$

Furthermore, since the sum of these two reduced masses is always one, it is possible to work with only one parameter μ. Usually, μ_2 is simply called μ in these applications, and m_1 then becomes $1 - \mu$. In this way μ is always between 0.0 and 0.5, since the second primary is the less massive.

Finally, the new unit of time is $1/\omega$. Normally, ω is expressed in a number of radians per second, hence in 1/sec. Its inverse $1/\omega$ is therefore expressed in seconds and may be interpreted as a unit of time.

Thanks to these new units, we only need to use one parameter to define the whole framework. The choice of parameter μ gives us the following data:

the mass of the first primary: $1 - \mu$,

the mass of the second primary: μ,

the coordinates of the first primary: $(\mu, 0)$, and

the coordinates of the second primary: $(\mu - 1, 0)$.

The coordinates of the primary follow from the fact that their distance is one combined with the relation for the center of gravity (3), and a minus for the second primary which is on the negative side of the X-axis is added. The new units will also modify the equations of motion of the third body and transform them into

$$x'' = -(1-\mu)\frac{x-\mu}{R_1^3} - \mu\frac{x+1-\mu}{R_2^3} + x + 2y' \tag{12}$$

and

$$y'' = -(1-\mu)\frac{y}{R_1^3} - \mu\frac{y}{R_2^3} + y - 2x' \tag{13}$$

with

$$R_1 = \sqrt{(x-\mu)^2 + y^2} \tag{14}$$

and

$$R_2 = \sqrt{(x+1-\mu)^2 + y^2}. \tag{15}$$

These formulae are less complicated than the original ones, particularly since the primaries have fixed positions. We will use (12–15) in our program to compute the orbit of the third body. The only free parameter is the mass parameter μ. Some special values are as follows:

0.0121409319...	Earth-Moon system as primaries
0.000953875...	Sun-Jupiter system
0.5	The so-called Copenhagen problem (two equal masses)
0.0385208965...	Critical value of the mass parameter

- *Example:* Describe the general properties of a restricted three-body problem of a binary star with components of 20 and 5 solar masses and a period of 241.72 days.

The mass parameter $\mu = 5/(20+5) = 0.2$.

For the first primary: reduced mass $= 0.8$; position in the co-rotating axes: (0.2, 0).

For the second primary: reduced mass $= 0.2$; position: $(-0.8, 0)$.

The unit of mass is 25 solar masses $= 5 \times 10^{34}$ g.

The angular velocity: $\omega = 2\pi/P = 3.0085244 \times 10^{-7}$ rad/sec.

The unit of time $= 1/\omega = 3.3239 \times 10^{6}$ sec.

The unit of length may be derived from Kepler's law: $G(m_1 + m_2) = A^3\omega^2$, which gives 3.328×10^{13} cm $= 2.2247$ A.U.

4.4 Numerical Method

The equations of motion given in (12,13) are differential equations of the second order. Since we have two of these equations, our problem is of the fourth order. This situation implies that we should select four initial conditions to start the numerical iteration procedure. These four are the initial position of the third body (x_o, y_o) and its initial velocity components (x'_o, y'_o). As the independent variable (the time) does not appear explicitly in the equations, the initial value of the time is unimportant and may be considered zero. The general form of the equations is

$$x'' = f(x, y, x', y')$$

and

$$y'' = g(x, y, x', y').$$

This system of two second-order equations may be transformed into a system of four first-order equations by treating the two velocity components as unknown functions of the time. We will therefore call these components u and v instead of x' and y'. Since the velocity component in a certain coordinate direction is the derivative of that coordinate, the equations then become

$$x' = u \tag{16}$$

$$y' = v \tag{17}$$

$$u' = -(1-\mu)\frac{x-\mu}{R_1^3} - \mu\frac{x+1-\mu}{R_2^3} + x + 2v \tag{18}$$

$$v' = -(1-\mu)\frac{y}{R_1^3} - \mu\frac{y}{R_2^3} + y - 2u \tag{19}$$

with

$$R_1 = \sqrt{(x-\mu)^2 + y^2}$$

and

$$R_2 = \sqrt{(x+1-\mu)^2 + y^2}.$$

This system may be more compactly written as

$$x' = fx;\ \ y' = fy;\ \ u' = fu;\ \ v' = fv \tag{20}$$

where fx, fy, fu, and fv are the right-hand sides of (16–19).

The problem is now to solve the equations of motion given by (16–19). This will be performed by numerical integration of these four coupled differential equations by means of the midpoint method. The iteration will be started from the four initial conditions:

- the initial positions (x_o, y_o), and
- the initial velocity components $(u_o = x'_o, v_o = y'_o)$.

Furthermore, a value of the mass parameter μ has to be selected in order to define the framework of the problem.

The way to compute the new state $(x_{i+1},\ y_{i+1},\ u_{i+1},\ v_{i+1})$ at time t_{i+1} from the state $(x_i,\ y_i,\ u_i,\ v_i)$ at time t_i is as follows. First, select a time step, Dt. Then

$$x_{i+1/2} = x_i + 0.5Dtf_x(x_, y_i, u_i, v_i) \tag{21a}$$

$$y_{i+1/2} = y_i + 0.5Dtf_y(x_i, y_i, u_i, v_i) \tag{21b}$$

$$u_{i+1/2} = u_i + 0.5Dtf_u(x_i, y_i, u_i, v_i) \tag{21c}$$

$$v_{i+1/2} = v_i + 0.5Dtf_v(x_i, y_i, u_i, v_i) \tag{21d}$$

and

$$x_{i+1} = x_i + Dtf_x\left(x_{i+1/2}, y_{i+1/2}, u_{i+1/2}, v_{i+1/2}\right) \tag{22a}$$

$$y_{i+1} = y_i + Dtf_y\left(x_{i+1/2}, y_{i+1/2}, u_{i+1/2}, v_{i+1/2}\right) \tag{22b}$$

$$u_{i+1} = u_i + Dtf_u\left(x_{i+1/2}, y_{i+1/2}, u_{i+1/2}, v_{i+1/2}\right) \tag{22c}$$

$$v_{i+1} = v_i + Dtf_v\left(x_{i+1/2}, y_{i+1/2}, u_{i+1/2}, v_{i+1/2}\right) \tag{22d}$$

in which the expressions for fx, fy, fu, and fv are given by the right-hand sides of equations (16–19).

In this way the new positions in the XY-plane and the corresponding velocity components at time t_{i+1} are obtained. Successive iterations of such points result in a series of positions in the orbital plane of the primaries (the XY-plane), representing the orbit of the third body relative to the positions of the primaries. The larger the times step Dt, the larger the distance between two successive positions of the third body. However, in order to insure the accuracy of the numerical procedure, the time step chosen should not be too large. Especially when the third body moves near one of the primaries, it is necessary to decrease the time step Dt to retain sufficient accuracy. Near the primaries the factor R_1 or R_2 becomes very small. Since these factors are found in the denominator of the equations of motion for the velocity components u and v, these two variables are very sensitive for small variations of R_1 and R_2 when these distances are small. Normally, a time step of 0.10 may be used when the third body is more than 0.5 away from each of the primaries. When it approaches one of the primaries closer than 0.5, the time step should be decreased to 0.02 or even 0.005 when closer than 0.25. There is no problem in changing the time step between two iterations. In any case Dt may not be changed during an iteration, i.e., between application of (21) and (22).

It is not possible to start an orbit from one of the primaries since, in that case, one of the distances would be zero. If one does not select the initial position too close to one of the primaries, initial velocity components between 0 and 1 may be taken for u_0 and v_0.

It is almost never possible to predict the orbit of the third body. The presence of the pseudo forces in the equations of motion (due to the co-rotating coordinate system) are mostly responsible for the (at first sight) strange behavior of the third body. Trying to direct the third body in an *a priori* chosen orbit is therefore very difficult and will succeed only after a number of trials. A change of 0.01 in one of the initial conditions of the equations often results in a very different orbit after a number of iterations.

4.5 Flowchart of the Program

The functions fx, fy, fu, and fv containing the equations of motion should be placed in a subprogram if possible since they are used twice each iteration. Furthermore, the distances R_1 and R_2 should be computed at the beginning of that subprogram since both R_1 and R_2 are needed in both the equations for f_u and f_v. The flowchart itself is rather simple:

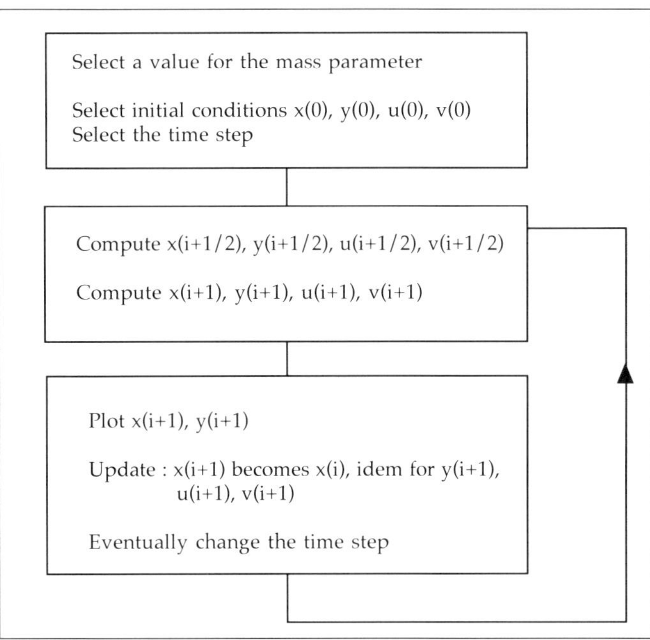

The program may be continued as long as the user desires. However, the cumulative nature of the small numerical and rounding errors will make the results less reliable as the number of iterations increases. Especially when the third body has passed very near to one of the primaries, the rest of the orbit becomes unsure.

4.6 Application: The Trojan Asteroids

The restricted three-body problem has many applications and the procedure described in this chapter can be used to compute a large variety of orbits by selecting appropriate initial conditions and mass parameters. If your computer has graphics or a plotter, you may draw the orbit immediately on your screen or listing. To test your program, we will now present a realistic application of the third body problem in the solar system.

The XY-plane of the restricted three-body problem contains five very special points, the so-called Lagrangian points. Detailed information on these points is given in the chapter on the equipotential surfaces of the two-body

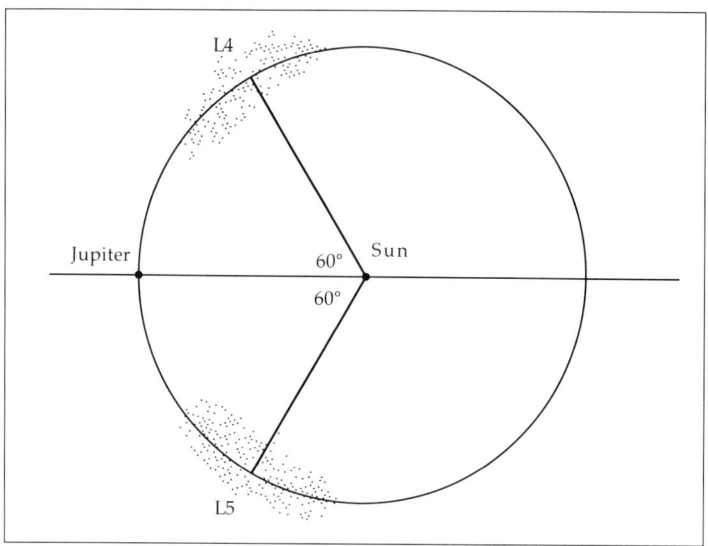

Fig. 4.3: *The Trojan asteroids are clustered in two clouds in Jupiter's orbit.*

problem. The first of these points, L_1 is located on the X-axis, somewhere between the two primaries, and always closer to the second primary than to the most massive one. This point is very important in the evolution of close binary stars. The two critical Roche lobes touch each other at L_1. At certain stages during the evolution of close binaries, mass transfer between the two components is possible when a component fills its Roche lobe. Two other Lagrangian points, L_2 and L_3, are also on the X-axis outside the positions of the primaries. The last two points, L_4 and L_5, are not located on the X-axis. They are situated at both sides of the X-axis in such a way that each of them forms an equilateral triangle with the two primaries. Therefore, they are also called the equilateral points. The exact positions of L_1, L_2, and L_3 may only be computed in an iterative way by numerical approximations. This will be explained in the chapter on the equipotential surfaces. On the other hand, L_4 and L_5 have trivial coordinates:

$$L_4 : \ (\mu - 0.5, \ 0.8660254\ldots)$$

$$L_5 : \ (\mu - 0.5, \ -0.8660254\ldots)$$

The importance of these two Lagrangian points lies in the fact that they represent the positions of two absolute extremes of the potential energy function $V(x, y)$ of the gravitational field of the primaries in the co-rotating coordinate system. Both are points where the gravitational force of the two primaries and the pseudo force of the co-rotating system balance so that the resulting force vanishes. A point-like mass with zero initial velocity placed in one of the equilateral points will stay there without moving. The other three points on the X-axis are saddle points. In the direction of the X-axis, they are minima; in the direction of the Y-axis, maxima.

When scientists began to explore the nature of the potential energy function and its effect on the motion of the third body, they discovered that stable periodic orbits around the equilateral points are possible if the mass parameter is smaller than the critical value of 0.03852.... They have also constructed methods to select consistent initial conditions for such periodic orbits. V. Szebehely presents an excellent and detailed overview on the characteristics of these orbits and on the problems and conditions for periodic orbits around the three co-linear points in *Theory of Orbits*, his textbook on celestial mechanics.

The most interesting application of periodic orbits around the equilateral Lagrangian points is the existence of the Trojan asteroids. These groups of asteroids revolve around the Sun in the same orbit and with the same orbital velocity as Jupiter. Although the individual members of the groups have their own orbits around one of the equilateral Lagrangian points, the two groups as a whole are 60° before and after Jupiter in the same orbit.

The reason for this phenomenon is the fact that each of the two clouds is clustered around an equilateral point of a two-body system with the Sun and Jupiter as primaries. Each of the Trojans has its own periodic orbit and acts as a third body in this system. We expect such periodic orbits in the Sun-Jupiter system since its mass parameter of 0.000953875 is smaller than the critical value.

Consequently, we will consider hypothetical Trojan asteroids in the Sun-Jupiter system and compute their orbit with our program. It is therefore necessary to ignore the eccentricity of the orbit of Jupiter (0.048417) and to assume that the asteroid moves exactly in the orbital plane of the large planet. The computation of the orbit has to be performed very accurately in order to keep the precious equilibrium and stability of the orbit. Therefore, the initial conditions must be selected with great care. We will consider two orbits for which we have computed the initial conditions with a procedure given by E. Rabe in 1961. With this method, which can be used to compute a large number of initial conditions, we obtained the following conditions:

Orbit A:

$$
\begin{aligned}
x(o) &= -0.509046125 \\
y(o) &= 0.883345912 \\
u(o) &= 0.0258975212 \\
v(o) &= 0.0149272418
\end{aligned}
$$

Orbit B:

$$
\begin{aligned}
x(o) &= -0.524046125 \\
y(o) &= 0.909326674 \\
u(o) &= 0.0646761399 \\
v(o) &= 0.0367068277
\end{aligned}
$$

These orbits are found around L_4, but initial conditions may be selected to compute orbits surrounding both the equilateral points. Such orbits, considered in the co-rotating coordinate system, start closer to Jupiter and move over L_4 around the Sun, following more or less the orbit of Jupiter. They then pass under L_5 and approach Jupiter as close as when they started in the upper part. Then they move back, pass just above L_5, go around the Sun, and follow L_4 to the point where they started.

Figure 4.4 shows the results for the two orbits. Orbit A was computed with a time step of 0.4 during the largest part of the orbit. The total time for one revolution around the Lagrangian point is about 80.3 time units. Orbit B, also computed with a time step of a few times 0.1, has a period of about 118 time units. Only at the two edges, where the trajectory sharply turns back, should the time step be decreased. At these points the velocity components change direction over a rather short distance.

The actual time needed for one revolution is easily found using the conversion factor between the actual and the reduced time. A reduced time interval of Dt corresponds to a real time interval DT according to

$$Dt = \omega DT.$$

The mean angular velocity of Jupiter in its orbit around the Sun is 0.52977 radians per year, or 1.6788×10^{-8} radians per second. This means that a time interval $Dt = 1$ in reduced units corresponds to a real time interval of 5.9567×10^7 seconds or 1.8876 years. This is the time unit of our problem. The period of revolution around L_4 for asteroid A is therefore 151 years, and for B it is 223 years. The unit of length is the mean distance between the Sun and Jupiter, which is 5.203 A.U., or about 780×10^6 km. Finally, the unit of mass is the sum of the masses of the two primaries. Since the mass of Jupiter is negligible compared to the Sun, we may take 2×10^{33} g as mass unit.

The mass parameter of the Earth-Moon system is 0.0121409319, which is smaller than the critical value. This opens the possibility of periodic orbits around L_4 and L_5. We are able to compute such orbits, but in reality perturbations of the Sun and the large planets are too great to leave the orbits stable. An artificial satellite placed in a periodic orbit would deviate from its

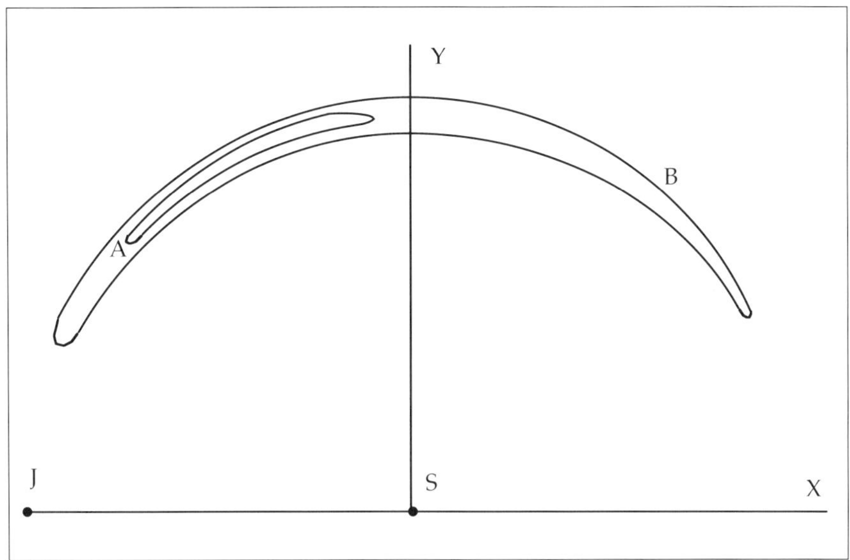

Fig. 4.4: *Orbits A and B of Trojan asteroid.*

ideal orbit. Orbits of this kind were studied by Rabe. We have applied his method to compute appropriate initial conditions.

For $\mu = 0.0121396054$, for instance, we found the following conditions:

Orbit C:

$$\begin{aligned} x(0) &= -0.4978603946 \\ y(0) &= 0.8833459119 \\ u(0) &= 0.0265752203 \\ v(0) &= 0.0146709149 \end{aligned}$$

Orbit D:

$$\begin{aligned} x(0) &= -0.5128603946 \\ y(0) &= 0.9093266740 \\ u(0) &= 0.0682722747 \\ v(0) &= 0.0334034039 \end{aligned}$$

In a binary star, with the two components acting as primaries, the attractive picture of a Trojan planet will never occur. To have such a situation the mass ratio of the two components should be as extreme as $1/25$, which is impossible. Mass ratios in main-sequence binary systems are normally never below 1/5.

As a test for your program, we present the first iterations of orbit A for the Trojan asteroid.

t	x	y	u	v
0.0	−0.509046125	0.883345912	0.0258975212	0.0149272418
0.4	−0.498648648	0.889247275	0.0260656072	0.0145894235
0.8	−0.488186942	0.895012259	0.0262198772	0.0142444218
1.2	−0.477666356	0.900637413	0.0263598894	0.0138887615
1.6	−0.467092970	0.906117740	0.0264824338	0.0135186022
2.0	−0.456474651	0.911446730	0.0265820581	0.0131306496
2.4	−0.445821802	0.916616747	0.0266519727	0.0127229632
2.8	−0.435147692	0.921619714	0.0266851922	0.0122955261
3.2	−0.424468301	0.926447959	0.0266757319	0.0118504843
3.6	−0.413801692	0.931095089	0.0266196653	0.0113920168
4.0	−0.403167012	0.935556741	0.0265158689	0.0109258604
4.4	−0.392583263	0.939831084	0.0263663286	0.0104585705
4.8	−0.382068047	0.943918980	0.0261759526	0.0099966414
5.2	−0.371636487	0.947823786	0.0259519135	0.0095456328
5.6	−0.361300488	0.951550817	0.0257026188	0.0091094470
6.0	−0.351068470	0.955106569	0.0254364741	0.0086898751
6.4	−0.340945619	0.958497819	0.0251606345	0.0082864852
6.8	−0.330934621	0.961730747	0.0248799468	0.0078968691
7.2	−0.321036752	0.964810233	0.0245962579	0.0075172034
7.6	−0.311253176	0.967739425	0.0243082031	0.0071430328
8.0	−0.301586210	0.970519651	0.0240115153	0.0067701418

The last few digits of these results may be different on your computer as a consequence of a smaller or larger number of significant digits. If your Trojan orbit smoothly joins the initial position from where you started, your program is correct.

4.7 Application: Orbit-Orbit Resonances

4.7.1 Nature of the Orbits

The Trojan asteroids are not the only cases whose orbital motion is mastered by Jupiter. The giant planet has captured other asteroids as well and really locked them up in a quite fixed orbit. This is the case with the asteroid Hilda and about twenty smaller companions forming the Hilda group, and with Thule, which seems to be alone. The coupling of the orbits of Jupiter and the asteroid is given by the ratio of the period of the asteroid to the period of Jupiter, P_A/P_J. In these cases this ratio may be written as a ratio of two small integers. For Hilda/Jupiter, it is 2/3; in the case of Thule/Jupiter, it is 3/4. In fact the Trojan asteroids may also be classified in this system with ratio 1/1. Similar situations are sometimes found between two members of the satellite systems of large planets (with the large planet acting as first primary) and between Pluto and Neptune (with the Sun as first primary), in which case the ratio is 3/2. Let us first focus on the case of Hilda.

The stability of the orbit of the asteroid is greatest when it passes through its perihelion just at the time when it crosses the radius vector of Jupiter. So, at that moment the Sun, Hilda, and Jupiter are on one straight line, and Hilda itself happens to be at its closest distance to the Sun. We will call this the "ideal" position. The motion of Hilda is, of course, mainly determined by the gravitation of the Sun. One can show that the gravitational effects of Jupiter over the orbit of Hilda cancel out so that after a while, we return to the situation in which the three are on one line, with Hilda in its perihelion. This occurs after three orbital revolutions of Hilda, during which Jupiter has orbited around the Sun two times. Figure 4.5 shows the stable orbit of Hilda, considered as third body in a restricted three-body problem with the Sun and Jupiter as primaries. We refer to subsection 4.7.2 for the initial conditions. If the position of the perihelion of Hilda is not exactly between the Sun and Jupiter, but makes an angle ω with the line between the two primaries, the orbit is also globally stable. After three orbits of Hilda, the asteroid reaches its perihelion at a smaller angle ω than where it started. The effect of Jupiter pulls the orbit in the direction of the ideal case. The resulting orbit is the typical triangle-like orbit, but now this triangle oscillates around the ideal position. In the case of Hilda, the amplitude of this oscillation, or to use the correct term, *libration*, is about 40°. The period of one libration is known to be 270 years.

4.7.2 Practical Examples

For the ideal orbit of Hilda, the stable nonlibrating triangle, one should start from

$$\begin{aligned} x_0 &= -0.647717531 \\ y_0 &= 0.0 \\ u_0 &= 0.0 \\ v_0 &= -0.6828143998. \end{aligned}$$

Using about 630 iterations with $Dt = 0.02$, a complete triangle taking three

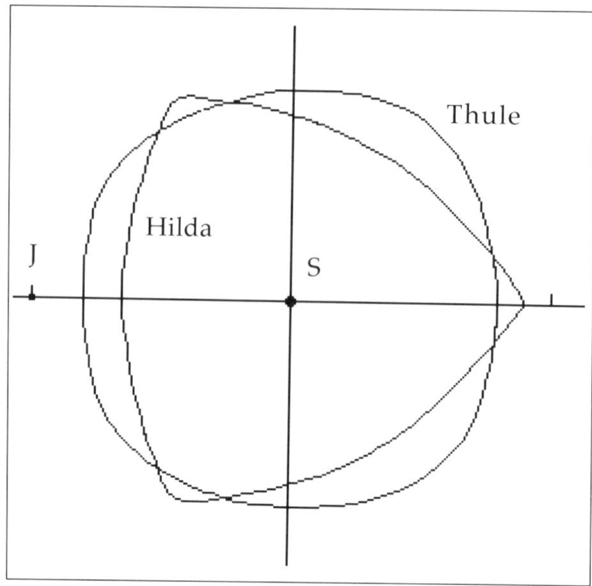

Fig. 4.5: *Ideal orbits of Hilda and Thule, without libration.*

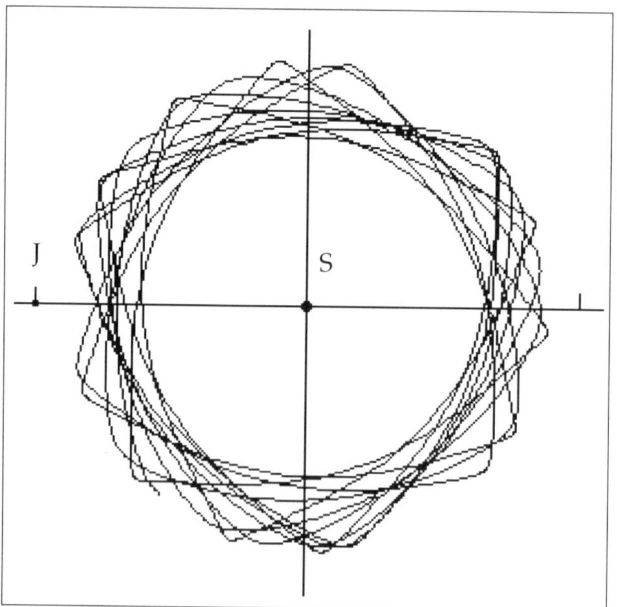

Fig. 4.6: *Orbit of Hilda with a libration of 40°.*

real revolutions of Hilda around the Sun in a fixed coordinate system is obtained.

For the more realistic case including the libration, one has to start from the following values:

$$\begin{aligned} x_0 &= -0.4952265404 \\ y_0 &= -0.4163448036 \\ u_0 &= 0.4389046359 \\ v_0 &= -0.5230661767. \end{aligned}$$

Now a larger number of iterations is needed. When we take a time step of $Dt = 0.05$, about 2800 iterations are needed to observe one libration (Fig. 4.6).

It was assumed in these calculations that the orbit of Jupiter is circular (in reality, $e = 0.048$) and that Hilda is moving in the orbital plane of the large planet (in reality there is an inclination of about 8°). Therefore, our results may deviate a little from the exact data.

In the case of Thule, the orbit has the shape of a square with the corners rounded off. The initial conditions are then

$$\begin{aligned} x_0 &= -0.7997634829 \\ y_0 &= 0.0 \\ u_0 &= 0.0 \\ v_0 &= -0.3334548184. \end{aligned}$$

They are given for the ideal case with no libration (Fig. 4.5).

4.7.3 The Case of Pluto/Neptune

This is the most important case in the solar system. It is well known that the orbit of Pluto has a large eccentricity of $e = 0.247$, which brings the planet at a certain moment inside the orbit of Neptune. The two planets are trapped in an orbit-orbit resonance. The period of Pluto is 3/2 times the period of Neptune. We have performed the calculations with the following orbital elements:

Neptune:

$$\begin{aligned} e &= 0 \\ P &= 165.62 \\ a &= 30.1584 \end{aligned}$$

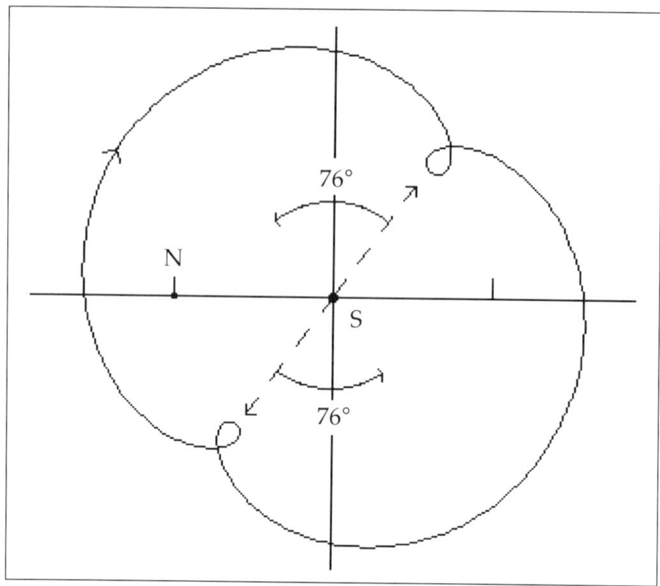

Fig. 4.7: *The basic part of the orbit of Pluto relative to the Sun and Neptune.*

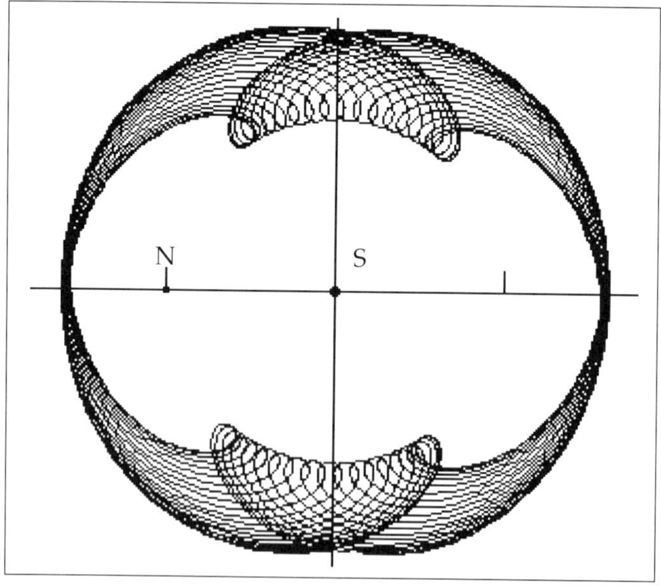

Fig. 4.8: *One libration taking about 20,000 years. The basic orbit of Fig. 4.7 oscillates with an amplitude of 38°.*

Pluto:

$$e = 0.247$$
$$P = 248.43$$
$$a = 39.5187$$

and μ : 0.0000525.

The period of Neptune is slightly overestimated in order to assure an exact resonance of 3/2. The amplitude of the libration is known to be 38°. The initial conditions are as follows:

$$x_0 = -0.6073955952$$
$$y_0 = -0.7774968265$$
$$u_0 = 0.1083342234$$
$$v_0 = -0.08463997159.$$

The typical shape of the orbit of Pluto in a co-rotating coordinate system with the Sun and Neptune as primaries is shown in Fig. 4.7. This part represents two revolutions of Pluto around the Sun. One can easily see that Pluto reaches two times a distance closer to the Sun than Neptune. The next two revolutions have the same shape, but the figure is rotated a little bit counterclockwise as the two perihelia approach the Y-axis. This phenomenon increases until the whole figure is rotated over 76°. This is half a libration, taking about 7,700 iterations with $Dt = 0.05$. Since the time unit in this case is 26.36 years, half a libration represents about 10,150 years. During the next 10,150 years the figure swings back to its initial position. At that point a whole libration has been performed. Computed with our program, a libration takes about 20,500 years. A total libration, shown in Fig. 4.8, illustrates that Pluto will never collide with Neptune since its distance to Neptune is always larger than about 17 A.U. (18 with professional results including the large inclination of the orbit of Pluto).

4.8 The Program Listing

The following sample program is written in MicroSoft QuickBasic 4.5:

```
DECLARE SUB effes (x, y, u, v, mu, fu, fv)

CLS
PRINT " "
PRINT " "
PRINT "Astrophysics With a PC : RESTRICTED THREEBODY PROBLEM"
PRINT "------------------------------------------------------------"
PRINT " "
PRINT "--------------- Minimal solution program ------------------"
```

```
PRINT "Input of initial conditions and parameters : ');"
PRINT ""
INPUT "Mass parameter  mu          : ", mu
INPUT "Initial conditions : x(o) : ", x
INPUT "                     y(o) : ", y
INPUT "                     u(o) : ", u
INPUT "                     v(o) : ", v
INPUT "Time step                 : ", dt

CLS
t = dt
n% = 20
ni% = 1

DO      'Main cycle computes blocks of 20 iterations and then asks the user
        'whether to continue or not

    CLS
    PRINT "    i        t            x            y            u            v "
    PRINT " "

'  next FOR-cycle computes 20 new iterations
    FOR i% = 1 TO n%

'  next block computes half step values of Cauchy method
        x1 = x + .5 * dt * u
        y1 = y + .5 * dt * v
        CALL effes(x, y, u, v, mu, fu, fv)
        u1 = u + .5 * dt * fu
        v1 = v + .5 * dt * fv

'  next block computes new state i+1 of Cauchy method
        x = x + dt * u1
        y = y + dt * v1
        CALL effes(x1, y1, u1, v1, mu, fu, fv)
        u = u + dt * fu
        v = v + dt * fv

'  new state is shown on screen
        PRINT USING "###### ######.#### ###.####### ###.#######_
 ###.####### ###.#######"; ni%; t; x; y; u; v
        t = t + dt
        ni% = ni% + 1

    NEXT i%

'  Ask user whether to continue or not
    PRINT "Continue (y/n) ?"
    DO
        ch$ = LCASE$(INKEY$)
    LOOP UNTIL (ch$ = "y") OR (ch$ = "n")
    PRINT " "

LOOP UNTIL ch$ = "n"

END

SUB effes (x, y, u, v, mu, fu, fv)
'
' computes the right hand sides (fu and fv) of the differential equations
' for the components of the velocity for input values of x,y,u and v, and mu
'
    r1 = SQR((x - mu) * (x - mu) + y * y)
    r2 = SQR((x + 1 - mu) * (x + 1 - mu) + y * y)
    fu = -(1 - mu) * (x - mu) / r1 ^ 3 - mu *_
 (x + 1 - mu) / r2 ^ 3 + x + 2 * v
    fv = -(1 - mu) * y / r1 ^ 3 - mu * y / r2 ^ 3 + y - 2 * u
END SUB
```

Chapter 5

Equipotential Surfaces of the Two-Body Problem

5.1 Introduction

We mentioned in the chapter on the restricted three-body problem the importance of the Lagrangian points in the potential energy field generated by the gravitation of two bodies (called primaries). The two equilateral points L_4 and L_5 were crucial for the understanding of the orbits of Trojan asteroids. This chapter deals with the specific properties of that potential energy field. The reader should first explore Chapter 4, which discusses the restricted three-body problem, before starting with this, since many of those concepts and the same coordinate system are used here.

An equipotential surface (EP-surface) around a massive body generating a strong gravitational field is the surface connecting and containing all points with equal potential energy. In case of a single spherical body, for instance, the equipotential surfaces are concentric spheres around the body. When rotational and heating effects are ignored, the terrestrial atmosphere is an example of this situation. Another example (at least as a first approximation) is the surface of the ocean adjusting itself to an equipotential surface of the Earth.

In case of a system with two bodies of comparable mass, for instance the two components of a binary star, the EP-surfaces are more complicated, not only because the gravitational field is generated at two places, but also because of the fact that the two bodies revolve around a common center of gravity. This second reason makes it necessary to study the problem in a coordinate system rotating with the bodies (cf. the restricted three-body problem), a situation which results in the appearance of pseudoforces in the expression for the potential energy.

We will only consider the potential energy in the orbital plane of two bodies having circular orbits. The intersection of an EP-surface with that orbital plane is then a curve connecting all the points $(x, y, 0)$ with the same equipotential energy. We will, therefore, use the expressions "equipotential surface" and "equipotential curve" or, shortly, "EP-surface" or "EP-curve" without any distinction. The shape of EP-surfaces in close binary systems is very important for their evolution since they offer a possible explanation for the formation of Wolf-Rayet stars in massive close binaries.

5.2 The Potential Energy Field

The potential field per unit of mass, generated by two bodies orbiting around their common center of gravity in a coordinate system rotating with the same angular velocity is given by

$$f(x, y) = -G\frac{M_1}{r_1} - G\frac{M_2}{r_2} - \omega^2\frac{x^2 + y^2}{2}. \qquad (1)$$

The variables in the above equation are defined as follows:

M_1 is the mass of the first primary. In the co-rotating coordinate system its coordinates are fixed at $(a, 0)$.

M_2 is the mass of the second primary, having coordinates $(-b, 0)$.

(x, y) is the coordinates of the point where the potential energy is computed.

ω is the angular velocity of the two bodies and the coordinate system.

r_1 is the distance from the point (x, y) to the first body, given by equation (2).

r_2 is the distance from the point (x, y) to the second body, given by equation (3).

$$r_1 = \sqrt{(x-a)^2 + y^2} \qquad (2)$$

$$r_2 = \sqrt{(x+b)^2 + y^2} \qquad (3)$$

Since the origin of the coordinate system is the center of gravity,

$$aM_1 = bM_2. \qquad (4)$$

And according to Kepler's law,

$$G(M_1 + M_2) = (a + b)^3\omega^2. \tag{5}$$

We are now able to proceed with new units of length, mass, and time.
New unit of length:

$$a + b. \tag{6a}$$

New unit of mass:

$$M_1 + M_2. \tag{6b}$$

New unit of time:

$$\frac{1}{\omega}. \tag{6c}$$

In this way, the two primaries form a two-body problem with a total mass of magnitude 1, at a distance 1 from each other and with an orbital period of 2π time units. In these units the masses of the two primary bodies are

$$\mu_1 = \frac{M_1}{M_1 + M_2} \qquad \mu_2 = \frac{M_2}{M_1 + M_2}. \tag{7a}$$

Since the total mass in the new units is always 1, it is sufficient to fix the mass of one primary. Therefore, we simply call the mass of the second (i.e., the less massive) component μ so that the mass of the first is $1 - \mu$. In this way the mass parameter μ is always between 0 and 0.5. We may also fix the coordinates of the two components in the co-rotating reduced system, taking into account equation (4) and the fact that the distance between the two components equals 1:
First primary:

$$(\mu, 0). \tag{7b}$$

Second primary:

$$(\mu - 1, 0). \tag{7c}$$

The potential energy in these new units is

$$\begin{aligned} f(x, y) = &-\frac{1 - \mu}{\sqrt{(x - \mu)^2 + y^2}} \\ &-\frac{\mu}{\sqrt{(x + 1 - \mu)^2 + y^2}} - \frac{x^2 + y^2}{2}. \end{aligned} \tag{8}$$

This function is negative for every point (x, y). A new function $V(x, y)$ is now introduced and given by

$$\begin{aligned} V(x, y) &= -f(x, y) \\ &= \frac{1-\mu}{\sqrt{(x-\mu)^2 + y^2}} + \frac{\mu}{\sqrt{(x+1-\mu)^2 + y^2}} + \frac{x^2 + y^2}{2}. \end{aligned} \tag{9}$$

An impression of $V(x, y)$ is given in Fig. 5.1.

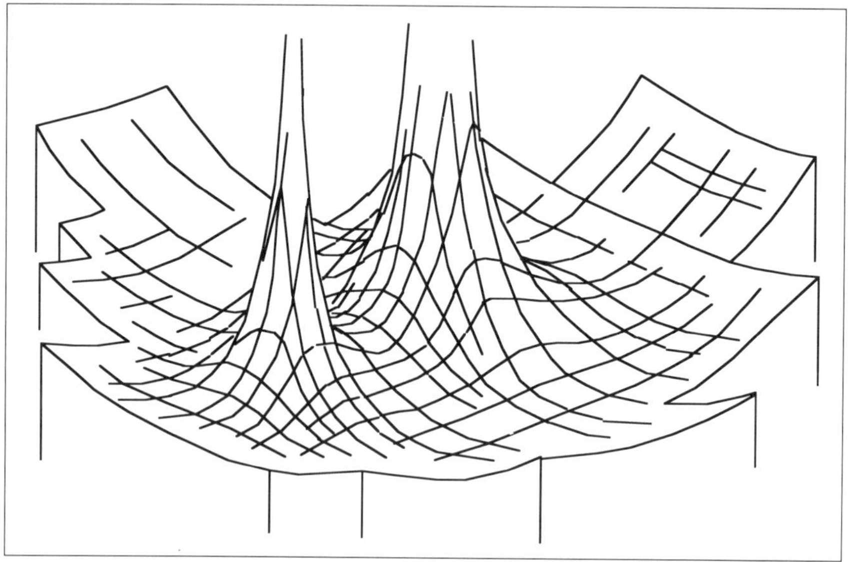

Fig. 5.1: *The potential function $V(x, y)$ of the circular two-body problem.*

The function $V(x, y)$ has a number of important characteristics:

- The function is symmetric in the coordinate y. If we replace y by $-y$ in the equation, the resulting function is the same. Therefore, only the part above the X-axis, the positive y-halfplane, is normally considered.
- $V(x, y)$ increases to infinity above the locations of the primaries located at $(\mu, 0)$ and $(\mu - 1, 0)$. Since both bodies are considered as point-like masses, the gravitation at their surface (i.e., at distance zero) is infinitely high.
- When x or y grows to infinity, the function $V(x, y)$ also grows to infinity as a consequence of the rotating coordinate system (cf. the last term in equation (9)).

- Above the X-axis, somewhere between the two primaries and at both sides, these functions have three saddle points. In the X-direction these points are minima of $V(x, y)$, but in the Y-direction they are maxima. These are the three collinear Lagrangian points L_1, L_2, and L_3. The first is the one between the primaries. L_2 is at the side of the less massive primary, and L_3 is at the side of the more massive.

- On the third edge of the equilateral triangle with the two primaries as first and second edge, there are absolute minima of $V(x, y)$. The minimum at the positive y-side is L_4; the other, L_5. These are the two equilateral Lagrangian points.

- The coordinates of the equilateral points are trivial since they are the edges of the triangles mentioned above:

$$L_4: \quad x = \mu - 0.5; \quad y = \sin 60^\circ = -0.8660254038$$
$$L_5: \quad x = \mu - 0.5; \quad y = -\sin 60^\circ = -0.8660254038$$

- The coordinates of the three collinear points on the X-axis are computed by numerical approximations. For every value of μ between 0.0 and 0.5, the function $V(x, y)$ in these points satisfies

$$V(L_3) \leq V(L_2) < V(L_1). \tag{10}$$

  The left inequality becomes equal for $\mu = 0.5$.

5.3 Computation of the Positions of the Collinear Points

Before calculating the locations of EP-curves, it is useful to compute first the positions of the collinear points, as they are linked to special cases of EP-curves. Since all three of them are on the X-axis, we limit the function $V(x, y)$ to its expression above X by putting y equal to zero in equation (9). Thus

$$VX(x) = V(x, y = 0). \tag{11}$$

An impression of $VX(x)$ for the case $\mu = 0.4$ is shown in Fig. 5.2. Putting $y = 0$ in (9) gives

$$VX(x) = \frac{1 - \mu}{\sqrt{(x - \mu)^2}} + \frac{\mu}{\sqrt{(x + 1 - \mu)^2}} + \frac{x^2}{2} \tag{12a}$$

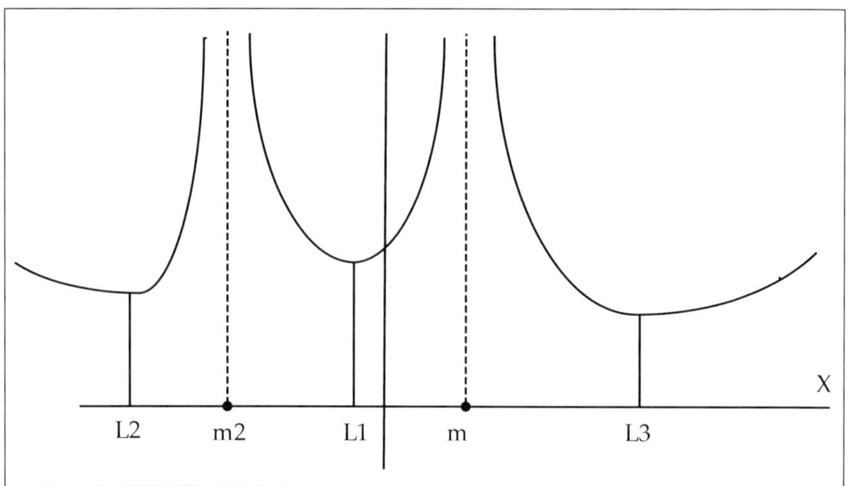

Fig. 5.2: *$VX(x)$ for $\mu = 0.4$.*

or in other words

$$VX(x) = \frac{1-\mu}{\text{abs}(x-\mu)} + \frac{\mu}{\text{abs}(x+1-\mu)} + \frac{x^2}{2}. \tag{12b}$$

The squares of the square roots in (12a) have been replaced here by absolute values. Let us then introduce sign parameters h_1 and h_2 to avoid the difficulties of having to work with absolute values during the numerical calculations. Thus

$$x - \mu = h_1 \, |x - \mu| \tag{13}$$

$$x + 1 - \mu = h_2 \, |x + 1 - \mu|. \tag{14}$$

So, we may write

$$VX(x) = h_1 \frac{1-\mu}{x-\mu} + h_2 \frac{\mu}{x+1-\mu} + \frac{x^2}{2}. \tag{15}$$

The values which should be taken for the sign parameters h_1 and h_2 depend on the place of the coordinate x on the X-axis.

- When $x > \mu$, the factor $x - \mu$ is positive, so $h_1 = 1$.
- When $x < \mu$, the factor $x - \mu$ is negative, so $h_1 = -1$.
- When $x > \mu - 1$, the factor $x + 1 - \mu$ is positive, so $h_2 = 1$.

- When $x < \mu - 1$, the factor $x + 1 - \mu$ is negative, so $h_2 = -1$.

Thus, we have the following three cases to consider:

$$\left.\begin{array}{l} x > \mu: \text{ so } h_1 = 1 \text{ and } h_2 = 1: L_3 \text{ is in this region.} \\ x < \mu \text{ and } x > \mu - 1: \text{ so } h_1 = -1 \text{ and } h_2 = 1: L_1 \text{ is in this region.} \\ x < \mu - 1: \text{ so } h_1 = -1 \text{ and } h_2 = -1: L_2 \text{ is in this region.} \end{array}\right] \quad (16)$$

These three cases cover all the possibilities. The right values of h_1 and h_2 should be inserted in (15) according to which of the three points one wants to compute. The rest of the formulae and the computational method remain unchanged for the three collinear points.

The positions of the three collinear points are under the three local minima of $VX(x)$. The problem is how to find a minimum of a function. This problem is the same as how to find a zero point of the derivative of that function, since in such a point the tangential line is horizontal and, therefore, the derivative of the function zero, in a minimum (or maximum). The derivative of $VX(x)$ is

$$VX'(x) = -h_1 \frac{1-\mu}{(x-\mu)^2} - h_2 \frac{\mu}{(x+1-\mu)^2} + x. \quad (17)$$

The zeros of this function may be found with the Newton-Raphson method described in the first chapter on numerical methods. This method may be applied successfully in this case since it is always possible to find appropriate starting values. We need the derivative of the function to use that method. Since our function is $VX'(x)$, the derivative to be applied in Newton-Raphson's iterative method is VX'', the second derivative of $VX(x)$. It is, therefore, obtained as the derivative of (17):

$$VX''(x) = 2h_1 \frac{1-\mu}{(x-\mu)^3} + 2h_2 \frac{\mu}{(x+1-\mu)^3} + 1. \quad (18)$$

The x-coordinate of one of the collinear Lagrangian points is then found by iteration on

$$x_{i+1} = x_i - \frac{VX'(x_i)}{VX''(x_i)}. \quad (19)$$

In both the expressions for VX' and VX'', the right values of h_1 and h_2 should be introduced according to the particular collinear point of which the position is computed. The next problem is to select good starting values for the iterations. The first collinear point L_1 always lies between the two primaries. Therefore, $x = 0.00$ is an appropriate starting value. The second collinear point L_2 is at the left side of the less massive primary. As its coordinate is thus certainly smaller than $\mu - 1$, we may start from $x = -1.00$. Finally, for L_3 at the right side of the more massive primary, a starting value $x = 1.00$ is a suitable choice. Once the positions of the collinear points have been obtained, the value of $VX(x, y)$ may be evaluated with (15).

5.3.1 Examples

a) Find the coordinates of L_1, L_2, and L_3 for $\mu = 0.40$.

L_1: $h_1 = -1$ and $h_2 = +1$ in the expressions for VX' and VX''. With $x(0) = 0.00$ as the starting value for (19), the next iterations are

$$\begin{aligned} x(1) &= -0.11251480 \\ x(2) &= -0.14155870 \\ x(3) &= -0.14161753 \\ x(4) &= -0.14161753. \end{aligned}$$

L_2: $h_1 = -1$ and $h_2 = -1$ in the expressions for VX' and VX''. With $x(0) = -1.00$ as the starting value for (19), the next iterations are

$$\begin{aligned} x(1) &= -1.12958896 \\ x(2) &= -1.21187328 \\ x(3) &= -1.23017691 \\ x(4) &= -1.23081306 \\ x(5) &= -1.23081377 \\ x(6) &= -1.23081377. \end{aligned}$$

L_3: $h_1 = +1$ and $h_2 = +1$ in the expressions for VX' and VX''. With $x(0) = 1.00$ as the starting value for (19), the next iterations are

$$\begin{aligned} x(1) &= 1.12189790 \\ x(2) &= 1.15970873 \\ x(3) &= 1.16203753 \\ x(4) &= 1.16204527 \\ x(5) &= 1.16204527. \end{aligned}$$

Thus we have the following values:

$$L_1 = -0.14162 \text{ and } VX(L_1) = 1.99045$$
$$L_2 = -1.23081 \text{ and } VX(L_2) = 1.75947$$
$$L_3 = \ \ 1.16205 \text{ and } VX(L_3) = 1.68954.$$

b) The table below lists the positions of the collinear points and their VX-values for a number of μ values.

μ	L_1	L_2	L_3	$VX(L_1)$	$VX(L_2)$	$VX(L_3)$
0.5000	−0.00000	−1.19841	1.19841	2.00000	1.72840	1.72840
0.4500	−0.07064	−1.21523	1.18061	1.99764	1.74513	1.70974
0.4000	−0.14162	−1.23081	1.16205	1.99045	1.75947	1.68954
0.3500	−0.21329	−1.24481	1.14287	1.97813	1.77077	1.66810
0.3000	−0.28613	−1.25673	1.12321	1.96007	1.77821	1.64568
0.2500	−0.36074	−1.26586	1.10317	1.93533	1.78060	1.62247
0.2000	−0.43808	−1.27105	1.08284	1.90233	1.77620	1.59866
0.1500	−0.51974	−1.27033	1.06230	1.85840	1.76218	1.57439
0.1000	−0.60904	−1.25970	1.04161	1.79848	1.73334	1.54979
0.0500	−0.71523	−1.22809	1.02083	1.71021	1.67720	1.52496
0.0300	−0.76964	−1.20119	1.01250	1.65895	1.63905	1.51499
0.0200	−0.80347	−1.18008	1.00833	1.62616	1.61287	1.51000
0.0100	−0.84808	−1.14677	1.00417	1.58382	1.57716	1.50500
0.0050	−0.88117	−1.11819	1.00208	1.55496	1.55163	1.50250
0.0030	−0.90037	−1.10029	1.00125	1.54002	1.53802	1.50150
0.0010	−0.93129	−1.06992	1.00042	1.51997	1.51931	1.50050
0.0005	−0.94549	−1.05553	1.00021	1.51280	1.51246	1.50025
0.0003	−0.95401	−1.04683	1.00012	1.50920	1.50990	1.50015
0.0001	−0.96807	−1.03243	1.00004	1.50449	1.50443	1.50005

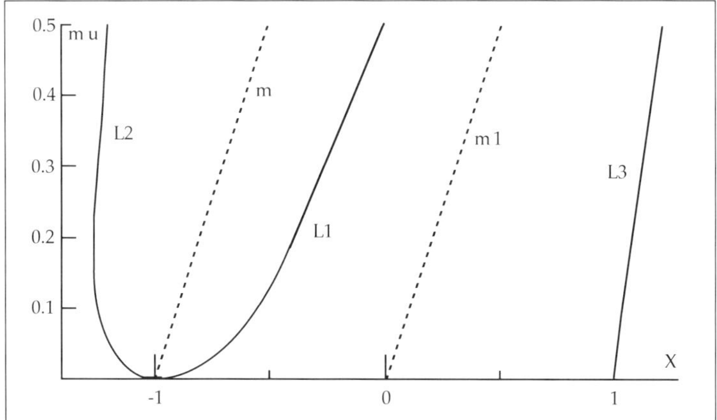

Fig. 5.3: *The positions of the collinear points for various μ.*

5.4 Computation of the Equipotential Curves

The intersection of an equipotential surface with the XY plane yields an equipotential curve (EC) in that plane. All points on such a curve have the same potential energy. The equation of any EC is simply

$$V(x, y) = K. \tag{20}$$

This defines the collection of points for which the potential energy is equal to a certain amount K. When a point (x, y) is on a specific EC, so is the point $(x, -y)$ because of the symmetry of $V(x, y)$ in the variable y. We therefore only consider positive y values. Selecting a value for K corresponds geometrically to an intersection of $V(x, y)$ with a horizontal plane at height K above the XY-plane of Fig. 5.1.

5.4.1 Qualitative Solution

Before actually computing the EC's, it is very interesting to study the problem first in a qualitative manner in order to see what shape the various EC's can have. We therefore use Figs. 5.1 and 5.2. Suppose we have selected a fixed value of μ.

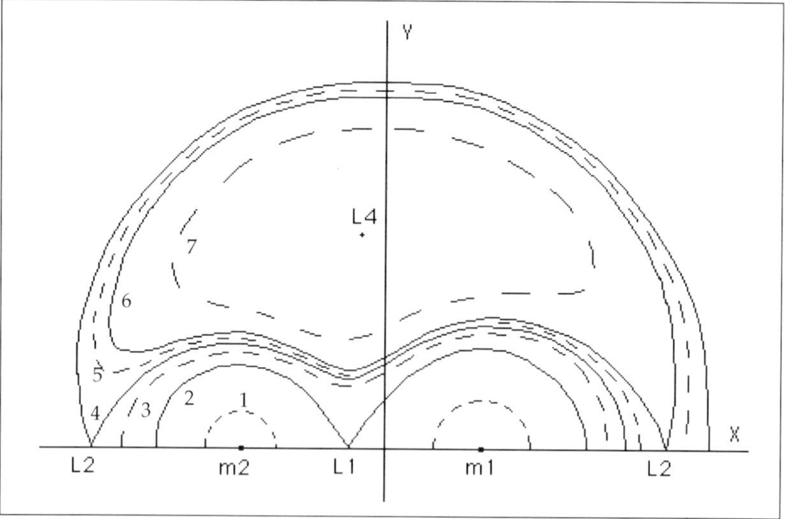

Fig. 5.4: *The seven types of equipotential curves. Solid lines (2,4,6) are passing through one of the collinear Lagrangian points. Dashed lines (1,3,5,7) are passing between them.*

We already mentioned the relation between the values of $V(x, y)$ in the five Lagrangian points:

$$V(L_4) = V(L_5) < V(L_3) \leq V(L_2) < V(L_1).$$

Since $V(x, y)$ has an absolute minimum in L_4 and L_5, there are seven types of equipotential curves (see Fig. 5.4) according to seven specific values or regions of K.

1. $K > V(L_1)$. We then take in Fig. 5.1 a horizontal intersection of $V(x, y)$ at a level exceeding the saddle point between the two primaries. Since $V(x, y)$ grows to infinity near the primaries and at large distances from the primaries, such an intersection consists of two quasi-circular parts around each primary and a large circle around the two primaries. The two small circles are due to gravitation, the large one to the pseudoforces induced by the co-rotation of the coordinate system.

2. $K = V(L_1)$. When K decreases from values larger than $V(L_1)$ to $V(L_1)$ itself, the two small circular parts around the primaries become larger while their shape deviates more and more from a circle. Both become stretched out in the direction of the other primary. When $K = V(L_1)$, they meet at L_1, the saddle point. This part of the EC is an "eight" balancing on the X-axis. The large circle of case (1) has become a little bit smaller but is still quasi-circular as it is mainly determined by the third term of the right-hand side of (9).

3. $V(L_2) < K < V(L_1)$. The large part of the EC becomes smaller but still encloses the two outer collinear points. Near the primaries, the EC consists of only one part containing the primaries and L_1. This curve, when placed in Fig. 5.1 is able to cross the saddle points of L_2 and L_3 but the saddle above L_1 has become too high.

4. $K = V(L_2)$. Both parts of the EC of case (3) now join above L_2. The solution now contains only one part. Near the third collinear point the solution crosses the X-axis twice. This EC is just able to cross the saddle in L_2.

5. $V(L_3) < K < V(L_2)$. The EC is one large curve. Near L_2, the EC is not able to cross the saddle, and it therefore turns back to swing around the two primaries all the way to the other side where it crosses the X-axis near L_3. The two intersections with the X-axis near L_3 approach each other as K decreases to the value $V(L_3)$.

6. $K = V(L_3)$. This is the last "special" curve. In L_3 both arms of the solution cross the X-axis at the same point, namely the saddle above L_3. This type of EC is the lowest capable of crossing the X-axis. The part of the EC in the lower half of the XY-plane is connected to the upper part at only one point.

7. $K < V(L_3)$. The EC's of this type never cross the X-axis because their potential energy is too low to climb the mountains and mountain passes above the X-axis. The EC now consists of two parts, one in each Y-halfplane. In the positive part, the EC is a closed curve around the absolute minimum L_4. The situation in the negative Y-halfplane is symmetrical but now around L_5. When K decreases to the value $V(L_4)$, the oval shape shrinks to a point at L_4. If K is taken smaller than $V(L_4)$ the solution is null.

From Fig. 5.4 one easily sees that the seven types can be classified according to their number of intersections with the X-axis.

Type:	1	2	3	4	5	6	7
Intersections:	6	5	4	3	2	1	0

5.4.2 Mathematical Solution

From a mathematical point of view, the problem is to compute the solution of the equation

$$V(x, y) = K \tag{21}$$

with V(x,y) given by (9) for a fixed value of the mass parameter μ and a number of K-values. This equation is a so-called implicit equation in x and y. It describes a relation between x and y, but cannot be written in the form $y = f(x)$ or $x = f(y)$ since it is not possible to eliminate one of the two variables completely. Each choice of K larger than the absolute minimum $V(L_4)$ gives a horizontal intersection of $V(x, y)$ and hence an equipotential curve. To find the position of such a curve, one may start in a certain point on it, and then actually walk along the EP-curve. Mathematically this is performed in the following way:

1. Select a point (x, y) and compute its function value $V(x, y)$. Let $K = V(x, y)$. This point is then a point on the EP-curve defined by (21).

2. Now suppose that $(x1, y1)$ is another point on the same EP-curve, close to our initial point; we may then write

$$x1 = x + Dx \tag{22}$$

$$y1 = y + Dy \tag{23}$$

with Dx and Dy as small quantities. We will now establish a relation between Dx and Dy in the case where the two points are on the same

EP-curve. Let us expand $V(x,y)$ in a Taylor series around initial point (x,y). We limit ourselves to the linear terms

$$V(x+Dx, y+Dy) = V(x,y) + \frac{\partial V}{\partial x}Dx + \frac{\partial V}{\partial y}Dy \tag{24}$$

in which $\partial V/\partial x$ and $\partial V/\partial y$ are the partial derivatives of $V(x,y)$. Their exact expression will be given later. This series is only an approximation to $V(x,y)$, but it is reasonable for small values of Dx and Dy. We may then combine (22,23,24) to obtain

$$V(x1, y1) = V(x,y) + \frac{\partial V}{\partial x}Dx + \frac{\partial V}{\partial y}Dy. \tag{25}$$

However, since both (x,y) and $(x1,y1)$ are on the same EP-curve, they both satisfy (21), which means that the V-factors in (25) can be omitted. The relation between Dx and Dy, the differences between two points on the same EP-curve, is then

$$\frac{\partial V}{\partial x}Dx + \frac{\partial V}{\partial y}Dy = 0. \tag{26}$$

Since Dx and Dy are small quantities, we may write their ratios as derivatives

$$y' = \frac{Dy}{Dx} \text{ and } x' = \frac{Dx}{Dy}. \tag{27}$$

Let us now introduce shorter notations: V'_x and V'_y instead of $\partial V/\partial x$ and $\partial V/\partial y$. Equation (26) may then be written in two ways:

$$y' = -\frac{V'_x}{V'_y} \tag{28}$$

and

$$x' = -\frac{V'_y}{V'_x}. \tag{29}$$

The expressions for V'_x and V'_y are (see formula 9)

$$V'_x = -\frac{(x-\mu)(1-\mu)}{\left((x-\mu)^2+y^2\right)^{3/2}} - \frac{\mu(x+1-\mu)}{\left((x+1-\mu)^2+y^2\right)^{3/2}} + x \tag{30}$$

and

$$V'_y = -\frac{y(1-\mu)}{\left((x-\mu)^2+y^2\right)^{3/2}} - \frac{\mu y}{\left((x+1-\mu)^2+y^2\right)^{3/2}} + y. \tag{31}$$

Thus we have two differential equations, (28) and (29), at our disposal to walk from one point on an EP-curve to another adjacent point on the same curve. Each of the differential equations has certain areas where it cannot be applied, but in most cases both may be used; therefore, one may switch from one equation to another without any problem. It is very easy to determine which equation to use in a certain case.

Figure 5.5 shows an example to explain which equation to use: in (28) y is written as a function of x. This equation can only be used in regions where the EP-curve is not parallel to the Y-axis. In the same way, (29)—giving x as a function of y—cannot be used when the EP-curve is more or less parallel to the X-axis.

An interesting method to decide whether to use (28) or (29) is based on the magnitude of Dx and Dy. Suppose that we are working with (28), i.e., with y considered as a function of x. This means in practice that we compute Dy for a certain step Dx. As soon as Dy becomes larger than Dx in absolute values, we are entering a region where the EP-curve makes an angle with the X-axis larger than 45° (or less than −45°) since the change of y is larger than the change of x. We then switch to (29) and proceed by considering x as a function of y. We have to use (29) until Dx becomes larger than Dy (always in absolute value), meaning that the angle has decreased below 45° (or above −45°). Then (28) is used again.

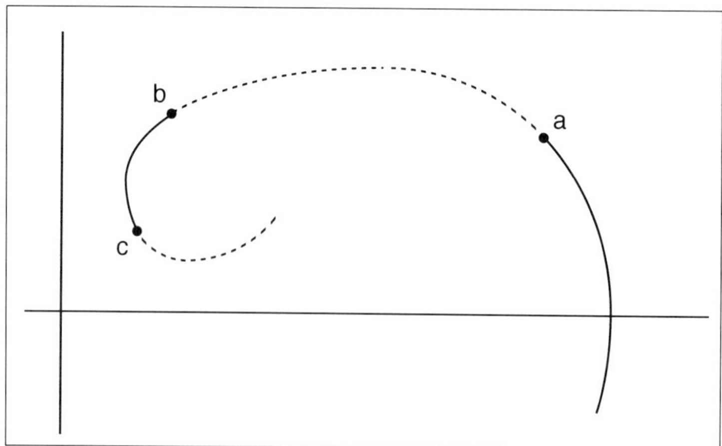

Fig. 5.5: *Application regions of (28), dashed lines; of (29), full lines.*

5.5 Numerical Method

Suppose we have to calculate the location of the EP-curve for a fixed value of the mass parameter μ. Let us take a certain point (x_i, y_i) and compute the corresponding potential energy $K = V(x_i, y_i)$. This point lies then on the EP-curve with equation $V(x, y) = K$. How may we proceed to another point labeled $i + 1$ on the same EP-curve close to the starting point? In the previous section, we learned how to walk along an EP-curve using (28) or (29).

1. If we use (28), then select Dx:

$$x_{i+1/2} = x_i + 0.5Dx \tag{32}$$

$$y_{i+1/2} = y_i - 0.5Dx \frac{V'_x(x_i, y_i)}{V'_y(x_i, y_i)} \tag{33}$$

$$x_{i+1} = x_i + Dx \tag{34}$$

$$y_{i+1} = y_i - Dx \frac{V'_x(x_{i+1/2}, y_{i+1/2})}{V'_y(x_{i+1/2}, y_{i+1/2})}. \tag{35}$$

2. If we use (29), then select Dy:

$$y_{i+1/2} = y_i + 0.5Dy \tag{36}$$

$$x_{i+1/2} = x_i - 0.5Dy \frac{V'_y(x_i, y_i)}{V'_x(x_i, y_i)} \tag{37}$$

$$y_{i+1} = y_i + Dy \tag{38}$$

$$x_{i+1} = x_i - Dy \frac{V'_y(x_{i+1/2}, y_{i+1/2})}{V'_x(x_{i+1/2}, y_{i+1/2})}. \tag{39}$$

It is obvious that both methods are very similar. The choice of which method to use depends on the magnitude of Dx and Dy. However it is very important to pay some attention to the sign of the step-size when switching from one case to another. Suppose we are proceeding along the EP-curve presented in Fig. 5.5, where we walk in the left side neighborhood of point a from right to left. We are then applying (28) with a negative step Dx. When we arrive at point b, it is time to switch to equation (29) since the slope of the EP-curve becomes too steep. This means that we now compute Dx for a fixed step Dy. If we continue with a positive step Dy, we would return along the part of the curve we have already computed. Since y was decreasing before point b, we have to continue with Dy negative in order to walk in the direction of point c.

When switching from method 1 (equation 28, y as a function of x) to method 2 (equation 29, x as a function of y), the new step-size Dy should be taken positive if y was increasing before the switch, and negative if y was decreasing. In the same way, when switching from (29) back to (28), the step-size Dx should be taken to be positive if x was increasing, and negative if x was decreasing. In this way, we will never go back on a part of the curve we have already passed along.

5.6 Some Remarks

1. Flowchart of the program:

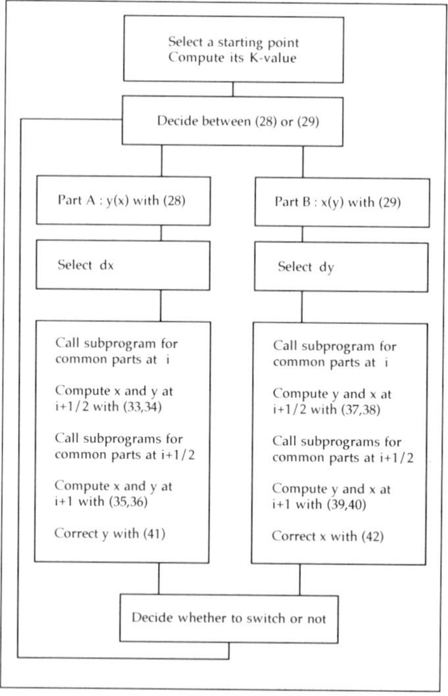

2. The computation of a set of EP-curves for a given value of μ requires a very large number of points. At least seven types of EP-curves should

be considered: the three with the K-values corresponding to the three Lagrangian points on the X-axis and four other curves between, above, and under these values. The two examples presented at the end of this chapter needed more than 1,000 points each.

3. Since all the points on an EP-curve satisfy the same relation

$$V(x, y) = K,$$

this relation may be used as a powerful tool to control the accuracy of the calculations. The successive points on an EP-curve are computed by numerical integration (32–39) and are therefore subject to errors. The more iterations we perform, the more our points will deviate from the exact EP-curve. This relation may be used to bring us back on the right track.

Assume, for instance, that we are calculating a certain EP-curve with $K = 2.1853$. After having computed a few points further on the curve, we become apprehensive about accuracy and decide to check the K-value of the last point computed. We therefore compute $V(x, y)$ and find 2.1779, which means that we are now on another, although close, EP-curve. In case we are using (28), y as a function of x, we may increase or decrease y by a small amount such as 0.001 and compute $V(x, y)$ again until we find the original K-value 2.1853. An easy and elegant way to perform such corrections is

$$y_{i+1}^{C} = y_{i+1} - \frac{V(x_{i+1}, y_{i+1}) - K}{V_y'(x_{i+1}, y_{i+1})} \tag{40}$$

with superscript C standing for the corrected value. This formula may be used several times until the numerator becomes zero. Each time then the corrected value y_{i+1}^{C} is used in the right-hand side to obtain a new and better correction. In case we are using (29), this correction procedure becomes

$$x_{i+1}^{C} = x_{i+1} - \frac{V(x_{i+1}, y_{i+1}) - K}{V_x'(x_{i+1}, y_{i+1})}. \tag{41}$$

If the correction is applied once after each new point is computed, we maintain very good accuracy no matter how many iterations are performed.

4. The program must contain instructions to compute (32–35) as well as (36–39). However, both parts are very similar and use a number of common parts. It is, therefore, best to compute these parts in a separate subprogram so that these formulae have to be programmed only once. Following are all the parts common to (28) and (29):

$$r_1 = \sqrt{(x-\mu)^2 + y^2}$$

$$r_2 = \sqrt{(x+1-\mu)^2 + y^2}$$

$$V'_x = -\frac{(x-\mu)(1-\mu)}{{r_1}^3} - \frac{\mu(x+1-\mu)}{{r_2}^3} + x$$

and

$$V'_y = -\frac{y(1-\mu)}{r_1^3} - \frac{\mu y}{r_2^3} + y.$$

These common parts are called twice during each iteration: a first time to evaluate the partial derivatives V'_x and V'_y in position i and a second time in $(i+1/2)$. The general structure of the program is shown in the flowchart diagram presented at the beginning of this section.

5. With the numerical procedure described above, it is not possible to start the iteration in an equilateral Lagrangian point, since in these points both the partial derivatives V'_x and V'_y are zero. Doing so would provoke overflow in (33, 35, 37, or 39). If we want to locate one of these three special EP-curves, we have to compute first the value $K = V(x,y)$ in the Lagrangian point. Next we search for a point in the XY-plane with the same K-value and then start the iteration to that point in both directions. Note also that some types of EP-curves consist of two separated parts. It is, of course, not possible to "jump" from one part to another.

5.7 Practical Examples

As a test for your program, we now present the first iterations of the EP-curve going through the point (1.05,0) for $\mu = 0.40$. The value of K for this point is 1.716751166. Since the EP-curves that cross the X-axis do so under a nearly right angle, we start with PART B in order to compute $x(y)$. A step-size of $Dy = 0.05$ is used.

y	x	$x_{(\text{corr})}$
0.00000	1.05000	1.05000
0.05000	1.04691	1.04691
0.10000	1.03763	1.03764
0.15000	1.02207	1.02208
0.20000	0.99999	1.00000
0.25000	0.97090	0.97090
0.30000	0.93391	0.93388
0.35000	0.88739	0.88729

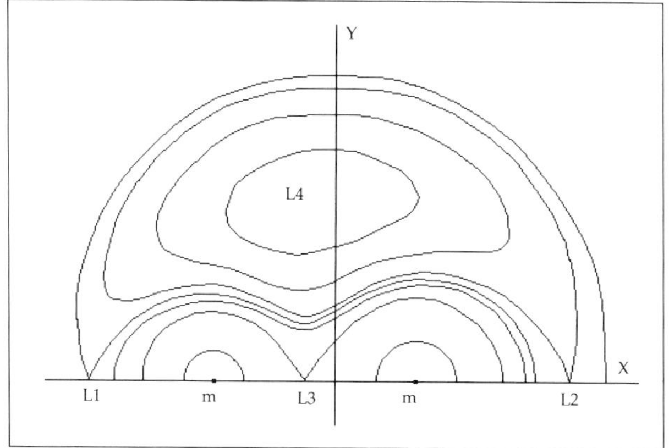

Fig. 5.6: *Equipotential curves for $\mu = 0.4$ computed with our program.*

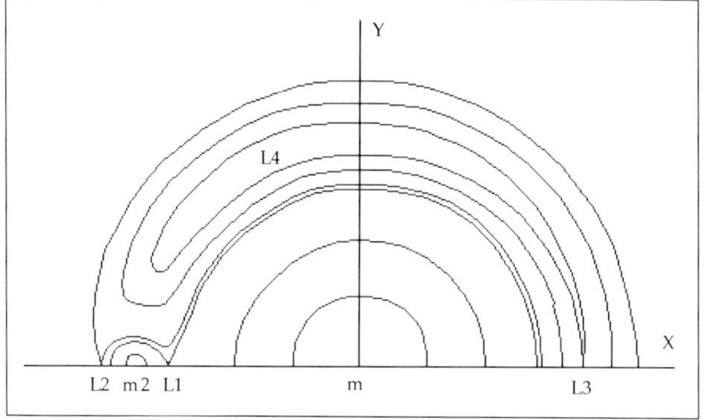

Fig. 5.7: *Equipotential curves for $\mu = 0.01$ computed with our program.*

The difference between two successive x-values has become larger than between to y-values. We switch to PART A (28) with $Dx = -0.05$ since x was decreasing.

x	y	$y_{(\text{corr})}$
0.83729	0.39256	0.39295
0.78729	0.42750	0.42773
0.73729	0.45582	0.45597
0.68729	0.47856	0.47866

For the last point, we obtain $V(x, y) = 1.716743872$, a difference of only 0.000007294 with the initial K-value. Without the corrections this would be

0.000782248, i.e., an error of a factor 100 larger than without a correction after each iteration.

5.8 Application: Close Binary Evolution

The equipotential surface passing through the first Lagrangian point L_1 is of great importance in the evolution of massive close binaries. The origin of stellar radiation is nuclear fusion. During the first ninety percent of the lifetime of a star, this fusion converts hydrogen into helium. This process takes place in the central region of the star. The star itself is very stable. There are no relevant changes in luminosity or radius. It is only when the central supply of hydrogen is exhausted, that drastic changes in the structure of the star occur. One of the consequences of hydrogen depletion is a rapid increase of the stellar radius. Of course, the word "rapid" should be seen in its astrophysical context—this process takes thousands of years, even for a massive star, but this is a very short period when compared with the total lifetime of a star.

An individual star has room to allow its radius to increase, and the same holds for wide binary stars. In the case of close binaries, however, the situation is different. As long as both components are in their core hydrogen-burning stage, they will behave like individual stars since the radii remain almost constant. When the more massive component of the binary, which evolves faster than the less massive secondary, has burned up its total central hydrogen supply, it will start increasing its radius. The stellar surface will always adjust itself to an equipotential surface. Therefore, the shape of the primary will deviate from its original spherical form and show a bulge in the direction of the secondary, just as the EP-surfaces do. At a certain moment, it will totally fill up the right part of the EP-surface passing through L_1 (see Fig. 5.8). We then say that the primary fills its *critical Roche lobe*. This fact, however, has no effect on the internal structure, so that the inner processes (not "knowing" what is happening outside) will continue to increase the radius of the star.

The stellar material at the surface will try to adjust itself to the shape of a larger equipotential surface. Any material exceeding the critical Roche lobe of the primary will then fall into the empty lobe of the secondary, just like an overflowing bucket. It will not fall directly on the surface of the secondary but first form an accretion disk in the orbital plane. A certain fraction may even escape from the system through L_2.

The consequences of the evolution of both components are, of course, very large since a considerable amount of matter may be transferred. For instance, a massive close binary with initial masses of $30M_\odot$ and $10M_\odot$ will start the mass transfer when the $30M_\odot$ component has reached core hydrogen depletion. The masses at that moment are about 26 and 10, since the initial

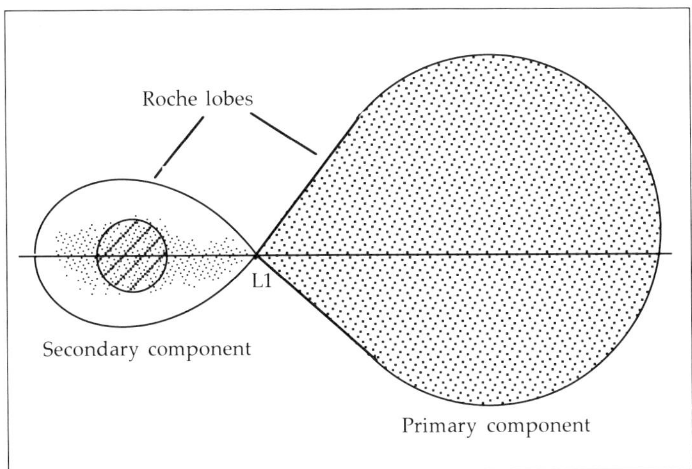

Fig. 5.8: *The more massive component of a close binary, filling its critical Roche lobe. Any further increase of the radius results in the transfer of mass from the primary to the secondary.*

$30M_\odot$ component has lost $4M_\odot$ due to stellar wind. About $14M_\odot$ will be transferred in about 10,000 years, resulting in a $12 + 24M_\odot$ system. The mass transfer will come to an end when the central temperature in the mass-losing star has become high enough to start helium burning. The new system consists of a massive $24M_\odot$ O-star, the former secondary, burning hydrogen in its core, and a smaller $12M_\odot$ component. This $12M_\odot$ star has a very unusual structure consisting of a convective core where helium is converted by nuclear fusion into carbon. Around it is an envelope with reduced hydrogen content since all the hydrogen rich layers have been lost during the process of mass transfer. Such a small but very bright, helium-rich star may be interpreted as a Wolf-Rayet star. A large number of these very special stars have been detected in close binaries as companions of more massive O-stars. The process of mass transfer through L_1 is a possible explanation for their origin, although individual Wolf-Rayet stars also exist and are obviously formed in another way—probably by the combined effects of stellar wind and convective mixing in a large part of the stellar interior.

The Wolf-Rayet component will evolve much faster than the O-star since the core helium burning stage takes only about 10% of the main sequence (= hydrogen burning) lifetime. The Wolf-Rayet star will suffer a high mass loss by stellar wind, but its final mass at the moment of helium depletion will still be large enough to exceed the minimal mass needed for a supernova explosion. The remnant after the explosion is a neutron star, a very compact, rapidly rotating object. It has the mass of a small star, but the radius of only a planet and a rotational period of only a few seconds. The large mass compared with the small radius means that the gravitational field in its neighborhood is extremely high. Part of the material lost by the O-star due to stellar wind will be attracted and accelerated as it falls into the strong gravitational field. The potential energy of these particles is radiated away as strong X-rays. Mass transfer is hence also an explanation for the massive X-ray binaries.

5.9 The Program Listing

The following sample program is written in MicroSoft QuickBasic 4.5:

```
DECLARE FUNCTION r1 (x, y, mu)
DECLARE FUNCTION r2 (x, y, mu)
DECLARE FUNCTION vxy (x, y, mu)
DECLARE FUNCTION dvdx (x, y, mu)
DECLARE FUNCTION dvdy (x, y, mu)


CLS
PRINT " Astrophysics With a PC : EQUIPOTENTIAL CURVES"
PRINT " ---------------------------------------------------"
PRINT " "
PRINT " --------------- Minimal solution program ----------------"
PRINT ""
PRINT " Input of initial conditions and parameters : "
INPUT "Mass parameter mu    : ", mu
INPUT "Starting point x(0) : ", x
INPUT "                y(0) : ", y


k = vxy(x, y, mu)
PRINT USING " Potential constant K = ####.#######"; k


' Ask user in which direction to move
PRINT "Enter 1 to compute Eq.curve as y(x), 2 to compute Eq.curve as x(y) "
DO
    drc$ = INKEY$
LOOP UNTIL (drc$ = "1") OR (drc$ = "2")


' select correct stepsize
IF drc$ = "1"
THEN
    INPUT "stepsize dx = ", dx
ELSE
    INPUT "stepsize dy = ", dy
END IF


' here starts main cycle
stp% = 0
DO


'   use midpoint method to compute y as function of x if drc% is 1

    IF drc$ = "1"
    THEN
```

```
        x12 = x + .5 * dx
        y12 = y - .5 * dx * dvdx(x, y, mu) / dvdy(x, y, mu)
        x = x + dx
        y = y - dx * dvdx(x12, y12, mu) / dvdy(x12, y12, mu)

'       add one correction of the computed value of y
        y = y - (vxy(x, y, mu) - k) / dvdy(x, y, mu)

'   use midpoint method to compute x as function of y if drc% is 2
    ELSE
        y12 = y + .5 * dy
        x12 = x - .5 * dy * dvdy(x, y, mu) / dvdx(x, y, mu)
        y = y + dy
        x = x - dy * dvdy(x12, y12, mu) / dvdx(x12, y12, mu)

'       add one correction of the computed value of x
        x = x - (vxy(x, y, mu) - k) / dvdx(x, y, mu)
    END IF

' show results on screen
    PRINT " "
    PRINT USING " x = ####.#######      y = ####.#######_
      K = ####.#######"; x; y; k
    PRINT " "

' ask the user what to do next
    PRINT " Enter c to change between x(y) and y(x)"
    PRINT " s to stop, or any key to continue with actual xy() or y(x)"
    DO
       chopt$ = INKEY$
    LOOP WHILE chopt$ = ""

' if user has answered with S : prepare to stop the program
    IF (chopt$ = "s") OR (chopt$ = "S") THEN stp% = 1

' if user has answered with C : change the direction and select new stepsize
    IF (chopt$ = "c") OR (chopt$ = "C")
    THEN
        IF drc$ = "1"
        THEN
            drc$ = "2"
            INPUT "stepsize dy = ", dy
        ELSE
            drc$ = "1"
            INPUT "stepsize dx = ", dx
        END IF
    END IF

    LOOP UNTIL stp% = 1

END


FUNCTION dvdx (x, y, mu)
'
' evaluates partial derivative dV/dx in (x,y)
'
    dvdx = -(1 - mu) * (x - mu) / (r1(x, y, mu)) ^ 3 - mu *_
 (x + 1 - mu) / (r2(x, y, mu) ^ 3) + x
```

```
END FUNCTION


FUNCTION dvdy (x, y, mu)
'
' evaluates partial derivative dV/dy in (x,y)
'
     dvdy = -(1 - mu) * y / (r1(x, y, mu) ^ 3)_
 - mu * y / (r2(x, y, mu) ^ 3) + y
END FUNCTION


FUNCTION r1 (x, y, mu)
'
' computes distance from point (x,y) to first primary
'
     r1 = SQR((x - mu) ^ 2 + y ^ 2)
END FUNCTION


FUNCTION r2 (x, y, mu)
'
' computes distance from point (x,y) to second primary
'
     r2 = SQR((x + 1 - mu) ^ 2 + y ^ 2)
END FUNCTION


FUNCTION vxy (x, y, mu)
'
' evaluates potential energy function V(x,y) at point (x,y)
'
    vxy = (1 - mu) / r1(x, y, mu) + mu / r2(x, y, mu) + (x ^ 2 + y ^ 2) / 2
END FUNCTION
```

Chapter 6

The Dynamical Parallax

6.1 Introduction

Binarity is a common phenomenon among stars. If a binary star has a sufficiently large period compared with its distance to the Earth, both components may be observed separately. Otherwise, binarity may be detected by means of spectroscopy. In the case that the two spectra are both visible, we will see each component shift periodically to shorter or longer wavelengths. Furthermore, if we look at the orbital plane of the binary star under a small angle, eclipses may be observed and allow us to derive information on the orbital elements. In some cases it is possible to estimate masses and radii of the two components. If only one of the spectra is observable, it is only possible to derive boundaries on the masses of the two components.

The dynamical parallax, a method to determine the masses and the distance of the binary, is only applicable to wide binaries. Both components should be observable separately. The masses and the distance to the Earth are then derived from the period, the apparent visual magnitudes, the bolometric corrections, and the apparent angle distance between the two components. All these quantities are observable for wide binaries. The method described here is limited to main sequence stars since we use the mass-luminosity relation for such stars. The method gives us the results by iteration of the parallax. A parallax computed in this way is called the *dynamical* or *hypothetical* parallax.

The masses and the parallax are computed by successive numerical approximations. Each new approximation is based on the results of the previous one. A similar method used to solve the equation for elliptic motion $E = M + e \sin E$ is presented in Chapter 1. The iterations are stopped when a certain accuracy is reached, for instance when a minimum number of significant digits in the results is no longer altered by new iterations. The number of iterations one needs is dependent on the characteristics of the problem, the method used to solve it and, of course, the accuracy demanded by the user. In our case, the method converges quite fast, and the number of approximations is never larger than ten.

6.2 Basic Formulae and Iterative Procedure

The computations are performed using some well-known formulae of basic astronomy. A subscript "1" is used for physical quantities referring to the first component (the primary) and a "2" for the secondary. All the logarithms have base ten.

The first formula is Kepler's law:

$$(M_1 + M_2)P^2 = A^3 = \frac{a^3}{p^3} \tag{1}$$

with

M_1: the mass of the primary expressed in solar units,

M_2: the mass of the secondary expressed in solar units,

P: the period expressed in years,

A: the semi-major axis of the binary orbit expressed in Astronomical Units (A.U.),

a: the apparent distance, expressed in arc seconds,

p: the parallax of the binary, expressed in arc seconds.

The formula relating the parallax and the magnitude of a star is

$$M_v = m_v + 5 + 5 \log p \tag{2}$$

with

M_v: the absolute visual magnitude,

m_v: the apparent visual magnitude.

This formula will be used for both components, and its variables will therefore be labeled "1" or "2". The same holds for the following formulae:

$$M_b = M_v - BC \tag{3}$$

with

BC as the bolometric correction, and

M_b as the bolometric magnitude.

The bolometric magnitude measures the total luminosity of the star, i.e., including radiation in all wavelength regions (X-ray, ultraviolet, visual, infrared...). The absolute visual magnitude measures only the radiation in the visual region. Therefore, bolometric magnitudes are smaller than visual magnitudes. Since a smaller magnitude number corresponds to a more luminous source, the bolometric correction is a positive quantity. Bolometric corrections are mainly dependent on the spectral type of the star and slightly on the luminosity class (supergiant, giant, main-sequence star...). For low mass stars, such as the Sun, BC is very close to zero since these stars produce most of their radiation at visual wavelengths. For very massive stars, the radiation distribution shifts to the blue so that bolometric corrections of a few magnitudes may occur. Bolometric corrections for special stars such as Wolf-Rayet stars are still very uncertain. A complete list of bolometric corrections may be found in most textbooks on stellar astronomy. The conversion of the absolute visual magnitude to the bolometric is needed to apply the next formula, the mass-luminosity relation

$$\log M = 0.58 - 0.112 M_b. \tag{4}$$

This relation is a good approximation and holds for masses larger than 0.6 solar masses ($M_\odot$) and for bolometric magnitudes between -10 and $+7$. Equation (4) can only be applied for main sequence stars. Other mass-luminosity relations have been determined for other classes of stars, such as giants or supergiants. Such relations are presented in the chapter on stellar atmospheres. Furthermore, the experienced reader may compute his own mass-luminosity relation for main sequence stars by means of the program described in the chapter on homogeneous stellar models.

The input parameters needed for the determination of the dynamical parallax are P, a, m_{v1}, m_{v2}, BC_1, and BC_2. The first four are obtained directly from observations. The two bolometric corrections may only be derived from a careful study of the spectra of each component. This study is used to check whether the two components are indeed main sequence stars so that (4) is valid.

We will now describe the subsequent steps of the iteration procedure. First, an initial guess of the masses has to be made to start the procedure.

a) Take

$$M_1 = M_2 = 1 M_\odot.$$

Other initial guesses are possible. The method is very stable so that poor initial guesses will only slow down the convergence.

b) The parallax may then be computed from (1) since the apparent distance a is observed. Then the two absolute visual magnitudes are obtained from (2) and the bolometric magnitudes from (3). Finally, these two bolometric magnitudes are used in (4) to calculate new estimates of the masses.

Step (b) is now repeated with these new approximations to obtain new estimates once again. This loop procedure is applied until the mass estimates have converged. A good criterion for convergence is asking that the masses are approximated with an accuracy of 0.01 $M_\odot$. Then the results are rounded off to 0.1 $M_\odot$.

Once the masses have been determined, the distance of the binary is computed using the last value of the parallax p. The distance expressed in parsec is then obtained from

$$d = \frac{1}{p}. \tag{5}$$

This formula is in fact the definition of an important unit of distance, the *parsec*. A distance of one parsec corresponds to a parallax of one arc second. Other units used for galactic and intergalactic distances are the Kpc (Kiloparsec = 1000 parsec) and the Mpc (Megaparsec = 10^6 parsec). The result of (5) has to be multiplied with 3.26 to express d in the more familiar light-years. A distance of one parsec corresponds to 206,265 A.U. (Astronomical Units); this quantity is also the number of arc seconds in one radian.

6.3 Concerning the Program

A flowchart of the program is shown on the following page. Since formulae (2, 3, and 4) are used twice in each iteration, it is a good practice to compute them in a subprogram. This may save some memory steps on programmable pocket calculators. The subprogram is then called once for each component.

6.4 Practical Examples

a) η Cas. Observed data:

$$\begin{aligned} P &= 526 \text{ yrs.} \\ a &= 12.21'' \\ m_{v,1} &= 3.7 \\ m_{v,2} &= 7.4 \\ BC_1 &= 0.03 \\ BC_2 &= 0.67 \end{aligned}$$

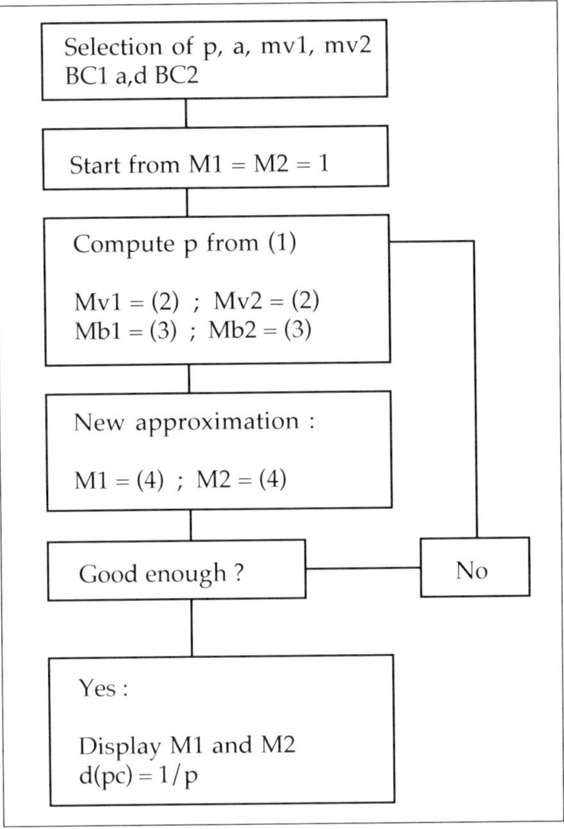

Starting from both masses equal to one we find the following:

$$p = 0.149''$$
$$M_{v,1} = 4.56$$
$$M_{v,2} = 8.26$$
$$M_{b,1} = 4.53$$
$$M_{b,2} = 7.59$$

hence:

$$M_1 = 1.18 \text{ and } M_2 = 0.54.$$

These approximations are then used for another iteration, which gives 1.15 and 0.52. The next iterations give 1.14 and 0.52 solar masses, which means that the method has converged. The parallax corresponding to these masses is 0.158″, which results in a distance of 6.3 parsec, or 20.6 light-years.

b) α Cen. Observed data:

$$\begin{aligned} P &= 78.8 \text{ yrs.} \\ a &= 17.6'' \\ m_{v,1} &= 0.3 \\ m_{v,2} &= 1.7 \\ BC_1 &= 0.06 \\ BC_2 &= 0.30 \end{aligned}$$

The iterations give the following data:

Iteration	M_1	M_2	p
1	1.00	1.00	0.760
2	1.15	0.85	0.761
3	1.15	0.85	0.761

Only two iterations are needed to obtain the results. The distance of this binary is 4.3 light-years. Apart from Proxima Centauri at a distance of only 4.27 light-years, α Cen is the nearest star to our Sun.

6.5 The Program Listing

The following sample program is written in MicroSoft QuickBasic 4.5:

```
CONST eps = .01

CLS
PRINT " "
PRINT " "
PRINT "Astrophysics With a PC : DYNAMICAL PARALLAX"
PRINT "-------------------------------------------"
PRINT " "
PRINT "------------- Minimal solution program ----------"
PRINT " "
PRINT "Input of the observed data :"
INPUT "Orbital period (years)                      : ", p
INPUT "Apparent distance (arc seconds)             : ", a
INPUT "Apparent magnitude of first component       : ", mv1
INPUT "Apparent magnitude of second component      : ", mv2
INPUT "Bolometric correction of first component    : ", bc1
INPUT "Bolometric correction of second component   : ", bc2
PRINT " "
PRINT "  i      m1        m2       dist      par       Mb1      Mb2"

' select starting values for the two masses
```

```
m1 = 1!
m2 = 1!
stopcrit% = 0
i% = 1

DO    ' Main cycle that stops when the two masses have converged

'   compute new approximations for the two masses
    par = a / p ^ (2 / 3) / (m1 + m2) ^ (1 / 3)
    mabs1 = mv1 + 5 + 5 * LOG(par) / LOG(10!)
    mabs2 = mv2 + 5 + 5 * LOG(par) / LOG(10!)
    mb1 = mabs1 - bc1
    mb2 = mabs2 - bc2
    m11 = 10 ^ (.58 - .112 * mb1)          ' = new approximation of first mass
    m22 = 10 ^ (.58 - .112 * mb2)          ' = new approximation of second mass
    dis = 1 / par * 3.26                   ' = new approximation of the distance

'   show iteration on screen
    PRINT USING "### ######.## ######.## #####.### ######.## ######.#_
 ######.#"; i%; m1; m2; par; dis; mb1; mb2

'   check convergence (if yes, stopcrit% becomes 1)
    IF (ABS(m1 - m11) < eps) AND (ABS(m2 - m22) < eps) THEN
        stopcrit% = 1
    ELSE
        m1 = m11
        m2 = m22
        i% = i% + 1
    END IF

LOOP UNTIL (stopcrit% = 1) OR (i% > 15)        ' End of main cycle

' show final results on screen
IF i% > 15 THEN
    PRINT "Method does not converge for your input"
ELSE
    PRINT USING "Final results : mass of 1st component : ###.##"; m11
    PRINT USING "                mass of 2nd component : ###.##"; m22
    PRINT USING "                distance in light yrs : ###.##"; dis
END IF

PRINT "Press any key to stop the program"
DO
LOOP WHILE INKEY$ = ""
END
```

Chapter 7

Polytropes

7.1 Introduction

Since we can only observe the surface of a star by studying the spectrum emerging from its atmosphere, we have to construct theoretical models of the stellar interior to get a better understanding of stellar structure. In practice, such a model is a large listing computed by a program containing a number of relevant physical quantities, such as the pressure, the temperature, the density, the chemical composition, and the radiation field, as a function of the distance from the center of the star or the amount of mass inside that distance. Furthermore, since the nuclear reactions in the central region of the star change the chemical composition, and since the energy released by these reactions is finally radiated away in space at the surface, the models are changing not only as a function of a space coordinate, but also in time. The evolution of a star is therefore computed as a number of successive models separated by certain time steps. Each of the models is constructed in agreement with the physical laws and as a logical evolution from the previous models. At a number of astrophysical institutes specializing in stellar structure and evolution, extended computer codes are developed to study the evolution of stellar models, including a large number of physical processes, such as all types of chemical mixing, detailed nucleosynthesis, and hydrodynamic effects. Until the early 1990s, the computation of these professional models required a large computer, due to the enormous amount of calculations involved. Only then were personal computers sufficiently powerful to allow the calculation of professional models to be performed in a reasonable time.

In this chapter we will describe a program that may be used to compute one of the most elementary stellar models, the *polytrope* or *polytrope star*. This model is only space dependent, as it gives the variation of the pressure, the density, and the mass as a function of the distance to the center of the model. No time dependent processes are included. The formulae have been developed by a number of well-known scientists such as Lane, Kelvin, Eddington, and Emden, who tried many decades ago to construct a simple stellar model. They knew that the two differential equations of hydrostatic equilibrium and of continuity of mass were not sufficient to determine the structure of a star in a unique way since these two first-order equations contain three unknown variables (mass, pressure, density) as function of the distance. Therefore, a third equation had to be found. This was the polytrope relation which connects the pressure and the density at a certain point in the star. This polytrope relation, characteristic of the models in this chapter, may be considered as the gas law for polytropes. It was selected more on mathematical than physical grounds since it cannot be deduced from physical arguments.

In fact, a polytrope is not a star but an enormous sphere of gas having a very specific type of equilibrium. It does not contain a number of important physical quantities and processes such as nuclear reactions, chemical mixing, energy transport, or mass loss by stellar wind; therefore, it has become history for professional stellar evolution. But most of all, it does not contain any information on the temperature throughout the star. We will study this type of stellar model for various reasons: it offers a good introduction on stellar structure and the construction of stellar models, it gives a good approximation of how certain physical parameters vary throughout the star, and it is the only stellar model that can be programmed on certain pocket calculators. A more sophisticated model—still based on polytropes but including nuclear reactions, energy transport, mixing, and information on the temperature—will be presented in another chapter. A home or personal computer is a must for that application. However, try first to understand and to program a polytrope, even if you have a computer in order to understand the larger model more easily.

The polytrope models contain one free parameter, the polytrope index, which enables us to consider models of various types.

7.2 Physical Background

In this section we will discuss all the relevant physical quantities and equations describing the polytrope model. The following notations are used:

M: the total mass of the star expressed in solar masses ($M_\odot$),

R: the total radius of the star expressed in solar radii ($R_\odot$),

r: the space coordinate giving the distance from the center of the star. At the center $r = 0$, and at the surface $r = R$,

$\rho(r)$: the density at distance r from the center,

$P(r)$: the pressure at distance r from the center,

ρ_c: the central density, equals $\rho(r = 0)$,

P_c: the central pressure, equals $P(r = 0)$,

ρ_m: the mean density, equals $M/\left(4/3\pi R^3\right)$,

$M(r)$: the amount of matter inside a sphere with radius r, so $M(0) = 0$ and $M(R) =$ the total mass M.

7.2.1 The Continuity of Mass

The following equation gives the variation of $M(r)$ as a function of r:

$$\frac{dM(r)}{dr} = 4\pi r^2 \rho(r). \tag{1}$$

$M(r)$ is the mass inside a sphere with radius r, so $dM(r)$ has the following meaning: consider a thin shell of thickness dr at distance r from the stellar center. The surface of that shell may be written $4\pi r^2$ so that its volume is approximated by $4\pi r^2 dr$. Since dr is very small, one may assume that the density is constant over the distance dr and may be taken as equal to the density at the inner boundary of the shell $\rho(r)$. Thus, the total mass of the shell is its volume multiplied by its density, $4\pi r^2 \rho(r) dr$, which is equal to $dM(r)$. Therefore, $dM(r)$ is the increase of mass at a certain radius r when that radius is increased by a small amount dr. In this way the star is divided into successive concentric thin shells from the center to the surface; $dM(r)$ is the mass inside such a shell. This equation is not used explicitly in the program, but it is incorporated in the final differential equation.

7.2.2 The Equation of Hydrostatic Equilibrium

This equation may be derived as follows. Consider a small vertical cylinder of matter in the stellar interior. The height of the cylinder is dr and its bottom and top surface is dA. As usual, dr and dA are very small when compared to the size of the star so that the physical variables such as the density may be considered constant in the cylinder. The star is said to be in hydrostatic equilibrium if such a cylinder always stays on the same place in the star. This means that the forces acting on the cylinder neutralize each other. Actually, three forces are working on the cylinder.

a) An upward pressure on the bottom surface. The force due to that pressure is equal to the product of the pressure and the surface on which it works. Thus, the magnitude of that force is

$$P(r)dA.$$

$P(r)$ is the pressure at the bottom of the cylinder, i.e., at distance r.

b) A downward pressure on the top of the cylinder. Since the top lies at a distance $r + dr$ from the stellar center, and since this force acts in the opposite direction (minus sign !) its magnitude is

$$-P(r + dr)dA.$$

c) The gravitational force acting downward on the cylinder. It is the product of the gravitational acceleration at distance r from the stellar center and the mass of the cylinder. The mass is the product of the volume and the density:

$$-G\frac{M(r)}{r^2}\rho(r)dr\ dA.$$

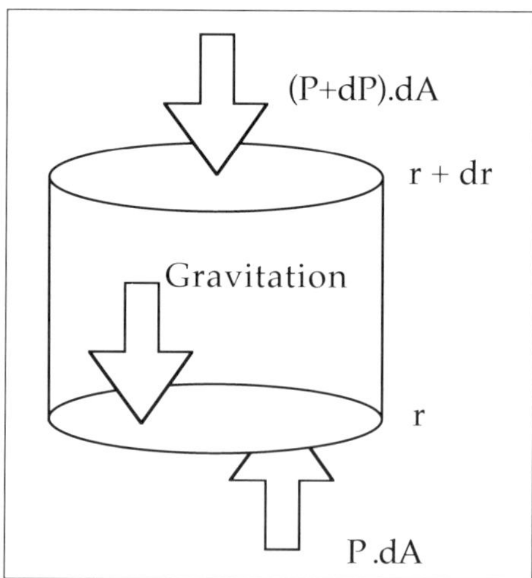

Fig. 7.1: *The three forces that balance hydrostatic equilibrium.*

Hydrostatic equilibrium means that the small cylinder stays at the same place, i.e., that the sum of all the forces acting on it is zero:

$$P(r)dA - P(r + dr)dA - G\frac{M(r)}{r^2}\rho(r)dr\ dA = 0.$$

When this equation is worked out using

$$P(r+dr) = P(r) + \frac{dP}{dr}dr,$$

the resulting equation is

$$\frac{dP}{dr} = -G\frac{M(r)}{r^2}\rho(r). \tag{2}$$

This is the very important equation of hydrostatic equilibrium which tells us that the gravitational force in the star is balanced by the variation of the pressure. When this equation is extended with velocity terms for a hydro-dynamic treatment, it is known as Euler's equation for hydrodynamic equilibrium. Equations of that kind may also be used, for instance, in models for planetary atmospheres. Equations (1) and (2) are first-order differential equations. Thus far we have three variables—$M(r)$, $P(r)$, and $\rho(r)$—which are functions of r, but only two equations. We will therefore complete our set with a third equation, the polytrope relation.

7.2.3 The Polytrope Relation

The following equation describes a relation between the density and the pressure:

$$P(r) = K\rho(r)^{(n+1)/n}.$$

K is constant throughout the star but may differ from one star to another. The free parameter n is the polytrope index, a positive real number between 0 and 5. Another form of this equation is

$$\frac{P(r)}{P_c} = \left(\frac{\rho(r)}{\rho_c}\right)^{(n+1)/n} \tag{3}$$

in which the mysterious constant K is replaced by the physically more meaningful central pressure and central density. The choice of n depends on the type of star one wants to study. The purpose of this chapter is, therefore, to compute the structure of the polytrope (i.e., the variation of the mass, the pressure, and the density as functions of r from the stellar center to the surface) for a given total mass and radius and for a given index n. To do so we will first merge the three equations of structure into one differential equation.

7.3 Mathematical Background

We will introduce an auxiliary function $F(r)$ defined as

$$F(r) = \left(\frac{\rho(r)}{\rho_c}\right)^{1/n}. \tag{4a}$$

Combining (3) and (4a) then gives

$$F(r) = \left(\frac{P(r)}{P_c}\right)^{1/(n+1)} \tag{4b}$$

or in other words

$$\rho(r) = \rho_c F^n \tag{4c}$$

and

$$P(r) = P_c F^{n+1}. \tag{4d}$$

We now start from the equation of hydrostatic equilibrium (2) which we write as

$$\frac{r^2}{\rho(r)} \times \frac{dP}{dr} = -GM(r).$$

With (4c) and (4d) this becomes

$$\frac{r^2(n+1)F^n P_c}{F^n \rho_c} \times \frac{dF}{dr} = -GM(r).$$

When both sides are differentiated to r and when (1) is substituted this becomes

$$F'' = -F^n \frac{4\pi G\rho_c^2}{(n+1)P_c} - 2\frac{F'}{r}$$

in which F'' and F' are the second and first derivatives of $F(r)$. We then define a new independent variable as

$$x = \frac{r}{r_n}$$

with the distance parameter r_n by:

$$r_n = \sqrt{\frac{(n+1)P_c}{4\pi G\rho_c^2}}. \tag{5}$$

So that finally

$$F'' = -F^n - 2\frac{F'}{x}. \tag{6}$$

The initial conditions of this equation are independent of the index n:

$$F(x = 0) = 1$$
$$F'(x = 0) = 0. \tag{7}$$

The point where $x = 0$ is also the point where $r = 0$, the stellar center.

Equation (6) is the Lane-Emden equation that describes the complete structure of a polytrope with index n. This equation will be solved to obtain $F(x)$, and then with (4c, 4d) the ratio $\rho(r)/\rho_c$ and P/P_c. In this way we only obtain the density and pressure relative to their central values, which are computed at the end. Also, r_n is only calculated at the end since it contains the central density and pressure (see eq. 5). This procedure will enable us to compute the central density and pressure for a given total mass and radius, two important characteristics with a clear physical meaning.

Let us first discuss some general properties of $F(x)$ since they are important for the interpretation of the results. We obtain for each value of the index n a different Lane-Emden equation, and therefore another solution $F(x)$, although the initial conditions are independent of n. For values of the polytrope index n between 0.0 and 5.0, $F(x)$ is a strictly decreasing function that crosses the X-axis at a certain zero point $x(n)$. The larger the value of n, the further the zero point from the origin. For $n = 0$, for instance, we have $x(n = 0) = 2.449$; for $n = 4$, this becomes $x(n = 4) = 14.9716$; and for $n = 5$, the zero point lies at infinity. For values larger than 5 there is no intersection with the X-axis. The physical meaning of $x(n)$ is clear since at that point $F(x)$ becomes zero. This means by (4c,4d) that the density and pressure are zero so that in reality we have reached the surface of the star. If n were larger than 5, there would be no zero point and the model would not have a surface. The density and pressure would then, after having reached a certain minimum, start increasing again, a physically meaningless situation. Therefore n must be selected between 0 and 5, but it doesn't have to be an integer. In the next section, we will solve (6) by iteration until $F(x)$ becomes negative. The exact value of $x(n)$ is therefore obtained only after having solved the equation from $x = 0$ to $x = x(n)$, a potentially lengthy process. For instance, using a step $dx = 0.1$ for a polytrope of index $n = 4$ would require 150 iterations. The same step for $n = 0$ needs only 25 iterations.

7.4 Numerical Procedure

The input parameters of the problem are the total mass M, the total radius R, and the polytrope index n. The problem is solved in four parts:

1. the first step from $x = 0$ to $x = dx$,

2. the other steps from $x = dx$ until $F(x)$ becomes negative,

3. the exact location of the zero point $x(n)$ where $F(x) = 0$, and

4. the determination of the central density and pressure.

Before tackling the Lane-Emden equation, we transform it into two first-order equations by considering the first derivative F' as a separate unknown function $H(x)$. Thus, we also have $F'' = H'$. The two equations are

$$F'(x) = H(x) \tag{8a}$$

and

$$H'(x) = -F(x)^n - 2\frac{H(x)}{x} \tag{8b}$$

with the initial conditions

$$F(x = 0) = 1 \text{ and } H(x = 0) = 0. \tag{8c}$$

Every solution starts at the Y-axis in $(0, 1)$ in a horizontal direction.

The numerical procedure to solve this problem, based on the midpoint method for two first-order equations, is then (with step-size dx)

$$x_{i+1/2} = x_i + 0.5\ dx \tag{9a}$$

$$F_{i+1/2} = F_i + 0.5\ dxH_i \tag{9b}$$

$$H_{i+1/2} = H_i + 0.5\ dx\left(-F_i^n - \frac{2}{x_i}H_i\right) \tag{9c}$$

and then

$$x_{i+1} = x_i + dx \tag{9d}$$

$$F_{i+1} = F_i + dx\ H_{i+1/2} \tag{9e}$$

$$H_{i+1} = H_i + dx\left(\left[-F_{i+1/2}\right]^n - \frac{2}{x_{i+1/2}}H_{i+1/2}\right). \tag{9f}$$

As usual, F_i stands for $F(x_i)$, and so on.

These formulae cannot be used for the first step from $x = 0$ to $x = dx$ since in that case we would have a division by zero in equation (9c). Computers generally don't handle division by zero well. To solve this problem the first step is computed in an alternative way to obtain the values of F and H for $i = 1$. We may use polynomial expansions around $x_0 = 0$ to compute the values at $x_1 = dx$. Such expansions are only valid in the close vicinity of the expansion point zero. Never use them with dx larger than 0.2 since then the error of the expansion, which is only an approximation, increases too much. The expressions for $F(x)$ and $H(x)$ near $x = 0$ are

$$F(x) = 1 - \frac{1}{6}x^2 + \frac{n}{120}x^4 \tag{10a}$$

and

$$H(x) = -\frac{1}{3}x + \frac{n}{30}x^3. \tag{10b}$$

So the first step simply becomes

$$x_1 = dx \tag{11a}$$

$$F_1 = 1 - \frac{1}{6}dx^2 + \frac{n}{120}dx^4 \tag{11b}$$

and

$$H_1 = -\frac{1}{3}dx + \frac{n}{30}dx^3. \tag{11c}$$

Since x is no longer zero, the normal iteration formula (9a–f) may be used. The first point computed with these formulae is x_2, starting from x_1. Once F_i is known, the values of the relative density and pressure at x_i are obtained from (4c,4d):

$$\frac{P}{P_c} = F_i^{n+1} \tag{12a}$$

$$\frac{\rho}{\rho_c} = F_i^n. \tag{12b}$$

Furthermore, the variation of the mass may be computed at each iteration from the quantity

$$LM(r) = -x_i^2 H_i. \tag{13}$$

The three unknown parameters ρ_c, P_c, and L will be calculated at the end when the exact location of the zero point $x(n)$ and the value of H in that point are known.

The next problem is the calculation of $x(n)$. During the iterations it is always necessary to check on both $F_{i+1/2}$ and F_{i+1}. If one of them is negative, the current iteration is immediately broken off. Do not attempt to finish the iteration, because when $F_{i+1/2}$ becomes negative, the calculation of H_{i+1} becomes impossible since this computation would need a real power of a negative quantity. The location of the zero point $x(n)$ and the value of H in that point are then obtained from the last point where $F(x)$ was still positive. Suppose that x_m is that last point, and let us call F_m and H_m the values of F and H in x_m. Then the values of H and x at the zero point are computed with sufficient accuracy from

$$x(n) = x_m - \frac{F_m}{H_m}. \tag{14}$$

The slope at $x(n)$ itself is also extrapolated from point x_m by

$$H(n) = H_m + \Big(x(n) - x_m\Big)\Big(-F_m^n - \frac{2}{x_m}H_m\Big). \tag{15}$$

Once $x(n)$ and $H(n)$ have been computed, the missing parameters are obtained from these results and from the total mass and radius selected by the user (both expressed in solar units):

$$P_c = \frac{9.048 \times 10^{14} M^2}{(n+1)H(n)^2 R^4} \text{ dyne/cm}^2 \tag{16a}$$

$$\rho_m = \frac{1.42M}{R^3} \text{ g/cm}^3 \tag{16b}$$

$$\rho_c = -\frac{\rho_m x(n)}{3H(n)} \text{ g/cm}^3 \tag{16c}$$

$$L = \frac{-x(n)^2 H(n)}{M} \tag{16d}$$

$$r_n = \frac{R}{x(n)}. \tag{16e}$$

This last parameter, the unit of length for that polytrope, is now expressed as R_o, but it may also be computed from its original expression (5). In that case it will be expressed in centimeters.

These five final expressions complete the theory on polytropes. For a given mass, radius, and polytrope index, the polytrope is computed as a table containing P/P_c, ρ/ρ_c, and $LM(r)$ as a function of x. At the end the relevant quantities are computed with (16a–e). Another approach would be to select the central density, the central pressure, and the index. We would then be able to compute the pressure, the density, and the mass immediately but would have to accept the total mass and radius as output parameters. Furthermore, it is easier to have a general idea on stellar masses and radii than on central densities and pressures. This is why we decided to follow the approach with the mass and radius as input parameters.

7.5 Some Remarks About the Program

- There is one fixed polytrope index for the whole model. It is also possible to construct models with more than one index corresponding to the physical state of the matter in certain regions of the star—for instance, convective motions in the interior of massive stars and radiative equilibrium in the outer regions. Such models are, of course, more complicated but also enable us to include important quantities such as the temperature and the luminosity in the model. A possible way to construct such composite polytrope models is discussed in another chapter of this monograph. However, first try to understand and to program a normal polytrope to get more familiar with the terminology and the pitfalls of the numerical method.

- It is better to print $\log P/P_c$ and $\log \rho/\rho_c$ in your program rather than P/P_c and ρ/ρ_c, since the latter become very small in the outer layers where the pressure and the density are very small compared to their central values.

- As we mentioned in the introduction to this chapter, polytropes are no longer used these days in professional astronomy. Following are some classical polytropes that were important earlier:

  $n = 1.5$ Model of a fully convective star where radiation pressure is not important; also used for a non-relativistic degenerate white dwarf.

  $n = 3$ The standard model of Eddington, used to describe a fully radiative star or a white dwarf with relativistic degeneration.

  $n = 5$ An approximation for the distribution of stars in a globular cluster. The model is then computed from zero to a certain distance since for $n = 5$, the zero point is at infinity.

- A flowchart of the program is shown on the following page. Try to use subprograms for the evaluation of the two differential equations. You will be able to call them twice during each step.

7.6 Applications

We first compute the values of $x(n)$ and $H(n)$ with our program for a number of polytropes indices and compare them with the exact results. These exact values are tabulated in almost every textbook on stellar structure. In this way they are used as a first test for the accuracy of our own program. Table 7-1 (see next page) lists the results for n between 0 and 3.5 together

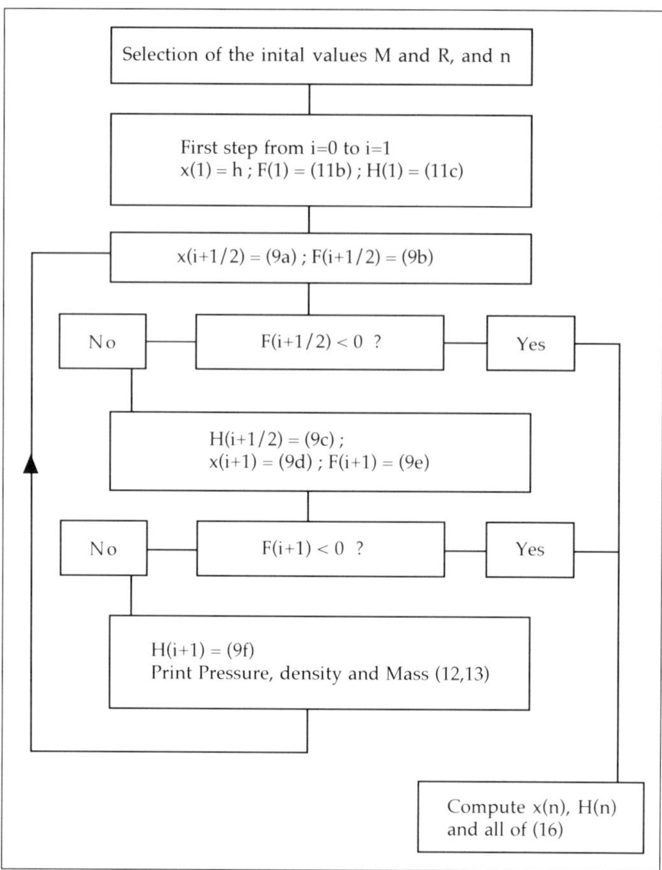

with the step-size used in our program and the number of steps needed to reach $x(n)$ with that step-size.

Table 7-1

n	h	iter.	$x(n)$	exact	$-x(n)^2H(n)$	exact
0.0	0.05	48	2.450	2.449	4.902	4.899
0.5	0.05	55	2.752	2.753	3.789	3.787
1.0	0.05	62	3.141	3.142	3.141	3.142
1.5	0.05	73	3.654	3.654	2.713	2.714
2.0	0.10	43	4.356	4.353	2.407	2.411
2.5	0.10	53	5.362	5.355	2.185	2.187
3.0	0.10	69	6.911	6.897	2.018	2.018
3.5	0.20	48	9.645	9.645	1.893	1.893

The error on $x(n)$ is never larger than about 1% for these values of the polytrope index, but this error probably increases for larger values of n. The relative error on $-x(n)^2H(n)$ is smaller, only a few times 0.1%. This shows that our method has a reasonable accuracy for the step-sizes used.

Secondly, we compute the structure of a polytrope with our program. The results presented in Table 7-2 (presented over the next two pages) are for $n = 1.5$ with a step-size of $h = 0.05$. We select a mass $M = 2$ and a radius $R = 3$.

The value $x = 3.65$ is the last one for which F is positive. We therefore compute $x(n)$ and $H(n)$ as follows:

$$x(n) = 3.65407$$

and

$$H(n)x(n)^2 = -2.7128.$$

When the mass and radius is taken into account one finds

$$\begin{aligned} P_c &= 4.330\ 10^{14}\ \text{dyne/cm}^2 \\ \rho_m &= 0.105185\ \text{g/cm}^3 \\ \rho_c &= 0.630583\ \text{g/cm}^3 \\ L &= 1.3564 \\ r_n &= 0.8210\ \text{solar radii}\ = 5.714 \times 10^{10}\text{cm}. \end{aligned}$$

Selecting another mass and radius will only affect these five values but not the solution of the polytrope.

Table 7-2

i	x	F	H	$\log P/P_c$	$\log \rho/\rho_c$	$LM(r)$
0	0.0000	1.00000	0.00000	0.0000	0.0000	−0.0000
1	0.0500	0.99958	−0.01666	−0.0005	−0.0003	0.0000
2	0.1000	0.99833	−0.03329	−0.0018	−0.0011	0.0003
3	0.1500	0.99626	−0.04983	−0.0041	−0.0024	0.0011
4	0.2000	0.99335	−0.06627	−0.0072	−0.0043	0.0027
5	0.2500	0.98963	−0.08256	−0.0113	−0.0068	0.0052
6	0.3000	0.98510	−0.09866	−0.0163	−0.0098	0.0089
7	0.3500	0.97976	−0.11455	−0.0222	−0.0133	0.0140
8	0.4000	0.97364	−0.13018	−0.0290	−0.0174	0.0208
9	0.4500	0.96675	−0.14552	−0.0367	−0.0220	0.0295
10	0.5000	0.95909	−0.16054	−0.0454	−0.0272	0.0401

(continued next page)

(continued from prior page)

i	x	F	H	$\log P/P_c$	$\log \rho/\rho_c$	$LM(r)$
11	0.5500	0.95069	−0.17522	−0.0549	−0.0329	0.0530
12	0.6000	0.94157	−0.18951	−0.0654	−0.0392	0.0682
13	0.6500	0.93174	−0.20340	−0.0768	−0.0461	0.0859
14	0.7000	0.92123	−0.21686	−0.0891	−0.0534	0.1063
15	0.7500	0.91005	−0.22985	−0.1023	−0.0614	0.1293
16	0.8000	0.89824	−0.24237	−0.1165	−0.0699	0.1551
17	0.8500	0.88582	−0.25438	−0.1316	−0.0790	0.1838
18	0.9000	0.87280	−0.26588	−0.1477	−0.0886	0.2154
19	0.9500	0.85923	−0.27683	−0.1647	−0.0988	0.2498
20	1.0000	0.84512	−0.28723	−0.1827	−0.1096	0.2872
21	1.0500	0.83051	−0.29707	−0.2016	−0.1210	0.3275
22	1.1000	0.81541	−0.30633	−0.2216	−0.1329	0.3707
23	1.1500	0.79987	−0.31500	−0.2424	−0.1455	0.4166
24	1.2000	0.78391	−0.32307	−0.2643	−0.1586	0.4652
25	1.2500	0.76757	−0.33055	−0.2872	−0.1723	0.5165
26	1.3000	0.75086	−0.33743	−0.3111	−0.1867	0.5703
27	1.3500	0.73382	−0.34370	−0.3360	−0.2016	0.6264
28	1.4000	0.71649	−0.34936	−0.3620	−0.2172	0.6848
29	1.4500	0.69889	−0.35443	−0.3890	−0.2334	0.7452
30	1.5000	0.68105	−0.35890	−0.4171	−0.2502	0.8075
31	1.5500	0.66300	−0.36278	−0.4462	−0.2677	0.8716
32	1.6000	0.64477	−0.36607	−0.4765	−0.2859	0.9371
33	1.6500	0.62639	−0.36879	−0.5079	−0.3047	1.0040
34	1.7000	0.60789	−0.37094	−0.5404	−0.3243	1.0720
35	1.7500	0.58930	−0.37254	−0.5742	−0.3445	1.1409
36	1.8000	0.57063	−0.37360	−0.6091	−0.3655	1.2105
37	1.8500	0.55194	−0.37413	−0.6453	−0.3872	1.2805
38	1.9000	0.53322	−0.37415	−0.6827	−0.4096	1.3507
39	1.9500	0.51452	−0.37368	−0.7215	−0.4329	1.4209
40	2.0000	0.49585	−0.37273	−0.7616	−0.4570	1.4909
41	2.0500	0.47725	−0.37132	−0.8031	−0.4819	1.5605
42	2.1000	0.45872	−0.36947	−0.8461	−0.5077	1.6294
43	2.1500	0.44030	−0.36720	−0.8906	−0.5344	1.6974
44	2.2000	0.42200	−0.36453	−0.9367	−0.5620	1.7643
45	2.2500	0.40385	−0.36148	−0.9845	−0.5907	1.8300
46	2.3000	0.38585	−0.35806	−1.0339	−0.6204	1.8941
47	2.3500	0.36804	−0.35431	−1.0853	−0.6512	1.9567
48	2.4000	0.35042	−0.35023	−1.1385	−0.6831	2.0173
49	2.4500	0.33302	−0.34586	−1.1938	−0.7163	2.0760
50	2.5000	0.31583	−0.34122	−1.2513	−0.7508	2.1326
51	2.5500	0.29889	−0.33632	−1.3112	−0.7867	2.1869
52	2.6000	0.28220	−0.33118	−1.3736	−0.8242	2.2388
53	2.6500	0.26577	−0.32583	−1.4387	−0.8632	2.2882
54	2.7000	0.24962	−0.32029	−1.5068	−0.9041	2.3349
55	2.7500	0.23375	−0.31458	−1.5781	−0.9469	2.3790
56	2.8000	0.21816	−0.30872	−1.6531	−0.9918	2.4203
57	2.8500	0.20287	−0.30272	−1.7319	−1.0392	2.4588
58	2.9000	0.18789	−0.29661	−1.8152	−1.0891	2.4945
59	2.9500	0.17321	−0.29041	−1.9036	−1.1421	2.5273
60	3.0000	0.15885	−0.28413	−1.9975	−1.1985	2.5572

(continued next page)

(continued from prior page)

i	x	F	H	$\log P/P_c$	$\log \rho/\rho_c$	$LM(r)$
61	3.0500	0.14480	−0.27780	−2.0981	−1.2589	2.5842
62	3.1000	0.13107	−0.27143	−2.2063	−1.3238	2.6084
63	3.1500	0.11766	−0.26504	−2.3235	−1.3941	2.6298
64	3.2000	0.10456	−0.25864	−2.4515	−1.4709	2.6485
65	3.2500	0.09179	−0.25225	−2.5930	−1.5558	2.6644
66	3.3000	0.07934	−0.24590	−2.7513	−1.6508	2.6778
67	3.3500	0.06720	−0.23959	−2.9315	−1.7589	2.6888
68	3.4000	0.05538	−0.23334	−3.1416	−1.8850	2.6974
69	3.4500	0.04387	−0.22717	−3.3946	−2.0368	2.7039
70	3.5000	0.03266	−0.22109	−3.7149	−2.2289	2.7084
71	3.5500	0.02176	−0.21513	−4.1559	−2.4936	2.7112
72	3.6000	0.01115	−0.20930	−4.8819	−2.9291	2.7125
73	3.6500	0.00083	−0.20363	−7.7043	−4.6226	2.7128

7.7 The Program Listing

The following sample program is written in MicroSoft QuickBasic 4.5:

```
DECLARE FUNCTION lg10 (x)

CONST g = 6.673E-08
      pi = 3.1415926536#

CLS
PRINT "Astrophysics With a PC : POLYTROPES"
PRINT "----------------------------------------"
PRINT " "
PRINT "-------- Minimal solution program -------"
PRINT " "
PRINT "Input of initial conditions and model parameters : "
INPUT "polytrope index : ", n
INPUT "stepsize        : ", dr
INPUT "mass            : ", mass
INPUT "radius          : ", rad

' show heading of main table of results on screen
PRINT "   i    x = r/rn       f           h       log(P/Pc)_
   log(d/dc)      L*mr"

' compute polytrope results in the center of the star
x = 0
f = 1
h = 0
i% = 0
p = f ^ (n + 1)
d = f ^ n
m = -x * x * h
PRINT USING " ### ####.#### #####.##### #####.##### ######.#### ######.####_
 ######.####"; i%; x; f; h; lg10(p); lg10(d); m

' compute first step
x = dr
f = 1! - x ^ 2 / 6 + x ^ 4 * n / 120
h = -x / 3 + x ^ 3 * n / 30
```

```
i% = 1
p = f ^ (n + 1)
d = f ^ n
m = -x * x * h
PRINT USING " ### ####.#### #####.##### #####.##### ######.#### ######.####_
 ######.####"; i%; x; f; h; lg10(p); lg10(d); m


' initialize main cycle
i% = 2
verder% = 1


DO
'  this is the main cycle

    x12 = x + .5 * dr
    f12 = f + .5 * dr * h

'  check whether  f12 (= F at half step) is still positive :
    IF f12 > 0!
    THEN
        h12 = h + .5 * dr * (-f ^ n - 2 * h / x)
        x1 = x + dr
        f1 = f + dr * h12
'      check whether  f1  (= F at new state) is still positive
        IF f1 > 0!
        THEN
            h1 = h + dr * (-f12 ^ n - 2 * h12 / x12)

'          compute pressure, density and mass
            p = f1 ^ (n + 1)
            d = f1 ^ n
            m = -x1 * x1 * h1

'          show results on screen
            PRINT USING " ### ####.#### #####.##### #####.##### ######.####_
 ######.#### ######.####"; i%; x; f; h; lg10(p); lg10(d); m

'          prepare for next iteration
            i% = i% + 1
            x = x1
            f = f1
            h = h1

'          pause every 15 iterations
            IF i% MOD 15 = 0
            THEN
                PRINT "Press any key to continue"
                DO
                LOOP WHILE INKEY$ = ""

            END IF
        ELSE
'         this ELSE is reached if f1 was negative
            verder% = 0
        END IF
    ELSE
'         this ELSE is reached if f12 was negative
        verder% = 0
    END IF

LOOP UNTIL verder% = 0    ' End of the main cycle


PRINT " "
PRINT "Press any key to proceed to the general characteristics"
DO
LOOP WHILE INKEY$ = ""
```

```
'compute general characteristics and surface data
xm = x - f / h
hm = h + (xm - x) * (-f ^ n - 2 * h / x)
fm = 0
pc = 9.048E+14 * mass * mass / (n + 1) / hm / hm / rad ^ 4
dm = 1.42 * mass / rad ^ 3
dc = -dm * xm / 3 / hm
lam = -xm * xm * hm / mass
rn = rad / xm

' show general characteristics and surface data on screen
PRINT " "
PRINT "central pressure (Pc) : "; pc
PRINT "average density       : "; dm
PRINT "central density (dc)  : "; dc
PRINT "mass parameter (L)    : "; lam
PRINT "distance unit (rn)    : "; rn
PRINT USING "x-final                  : #####.#### "; xm
PRINT USING "-x2*h (final)            : #####.#### "; -xm * xm * hm

PRINT " "
PRINT "Press return to stop the program"
DO
LOOP WHILE INKEY$ = ""
END


FUNCTION lg10 (x)
'
'  computes the base 10 logarithm of x using the natural logarithm LOG(x)
'
     lg10 = LOG(x) / LOG(10!)
END FUNCTION
```

Chapter 8

Homogeneous Stellar Models

8.1 Introduction

Before starting this chapter, the reader should first explore the chapter on polytropes because we will utilize the theory on polytropes and the mathematical techniques used to compute them. This chapter on homogeneous stellar models is one of the most complicated of this work and should only be attempted by experienced readers, who, we assume, are already familiar with the theory on polytropes.

The most simple model of a star is normally described by a system of four coupled differential equations and three algebraic equations. The four differential equations give the variation of M_r (i.e., the mass inside a sphere with radius r), the pressure P, the temperature T, and the luminosity L_r as functions of the radial distance r from the stellar center. The three algebraic equations are first, the equation of state, connecting P, T, the density, and the chemical composition; second, the equation for the opacity; and third, the equation for the nuclear energy production. The second and third are in professional programs mostly used in the form of large interpolation tables containing the opacity and energy production as functions of the density, the temperature, and the chemical composition.

The solution of this set of equations is rather difficult since two of the four differential equations have their boundary conditions at the center of the star (where M_r and L_r are zero) and two at the surface (where P and T have to join the P and T values of the stellar atmosphere, or in simplified models they may be considered zero). Therefore, it is not possible to solve the equations by iterations from the center to the surface or vice-versa since in both cases we have only two boundary conditions of the four we need to start the iterations.

Before starting with our alternative model, let us briefly describe how the boundary value problem is solved in professional programs. First, start from the center with estimates of the central pressure and temperature and compute the stellar structure outwards by iteration until a certain fraction of the mass is reached. This fraction is called *fitmass*. Then, start from the surface with estimates of the mass, radius, and the luminosity and compute the stellar structure inwards by iteration, down to the fitmass. At the fitmass, we now have two values for all the physical quantities (P, T, ...). If all the variables have a consistent pair of values, we have a preliminary model. If not, other (i.e., better) estimates at the center and the surface should be made by using numerical techniques based on methods for finding zeros of nonlinear functions. Once a preliminary model is obtained, it is further improved by numerical relaxation.

We will construct a simple model by fitting two polytropes, one for the convective core and another for the radiative envelope of the star. The method will only be suitable for stellar models with a homogeneous chemical composition, hence newly-born stars. These are the so-called Zero Age Main Sequence stars, or simply ZAMS-models. Although composite polytropes are less complicated than professional models, they reproduce some physical realities remarkably well.

We have mentioned in the chapter on polytropes that a polytrope index of 1.5 may be used to represent a totally convective star in the absence of radiation pressure. An index of 3 corresponds to a totally radiative star. Since massive stars have a convective core and a radiative envelope around them, it should be possible to start the iteration in the core with an index of 1.5 and to proceed outwards until the boundary of the core is reached. Then a new polytrope of index 3 is fitted to the core, and the iteration is continued outwards. Fitting two polytropes requires only a minimum of calculations. A possible method is given in *Stellar Structure* by S. Chandrasekhar. We will use an alternative method containing a free parameter that will be used to fit the model to the observations. The total number of free parameters is three: the central temperature, the central density, and the value of the outer polytrope function at the boundary of the core. However, our models should reproduce important observational facts such as the mass luminosity relation and the position of the main sequence in the HR diagram. With these restrictions the number of free parameters reduces to only one. For practical use we will choose the total mass of the star as the only free parameter because the total mass of a star is its most important physical property. Then we will express the three initial free parameters as functions of the total mass.

8.2 Physical Background

8.2.1 The Pressure, Density, and Mass

Two of the four basic equations of stellar structure have already been discussed in the chapter on polytropes: the equation of continuity of mass and the equation of hydrostatic equilibrium. These equations are

$$dM_r = 4\pi\rho r^2 dr \tag{1}$$

and

$$dP = -G\rho\frac{M_r}{r^2}dr. \tag{2}$$

Combination with the polytrope relation

$$P \sim \rho^{(n+1)/n} \tag{3}$$

results in the Lane-Emden equation:

$$F'' = -F^n - \frac{2F'}{x}.$$

This second-order equation is then transformed into a system of two first-order differential equations:

$$F' = H \quad \text{and} \quad H' = -F^n - \frac{2H}{x}. \tag{4}$$

When solved with the initial conditions

$$F = 1 \quad \text{and} \quad H = 0 \text{ at } x = 0 \tag{5}$$

they describe the structure of a polytrope of index n. The pressure, density mass, and distance are obtained from

$$P = P_c F^{n+1} \tag{6}$$

$$\rho = \rho_c F^n \tag{7}$$

$$M_r = -4\pi\rho_c {r_n}^3 x^2 H \tag{8}$$

and

$$r = r_n x \tag{9}$$

in which the index c points at the central values of the pressure and the density, and M_r is the amount of mass within a distance r from the center. The distance parameter r_n is given by

$$r_n = \sqrt{\frac{(n+1)P_c}{4\pi G {\rho_c}^2}} \tag{10}$$

where G is the gravitational constant.

In the chapter on polytropes, we have computed the pressure and density of the polytrope relative to the central values. Because of our decision to use the mass and the radius of the polytrope—rather than the central density and pressure—as the two free parameters, the central density and central pressure were computed only after the exact position of the surface of the polytrope was found.

In this application, however, we need to know the exact values of the central pressure and central density from the start. Only in this way will it be possible to compute the real absolute values of the physical quantities (P, T, ...) during the iterations needed to solve the polytrope. On the other hand, the mass and radius will, in principle, be known only at the end of the calculations when the surface of the star is reached. Knowing the exact values of pressure and density is necessary to compute the temperature from the equation of state and the nuclear energy production.

At this point we encounter the most significant difference between composite polytrope models and professional models. Since we compute the pressure and density from the polytrope equation and only afterwards, the temperature from the pressure and density, the temperature itself has no influence on the pressure and the density. In professional models, on the other hand, the temperature is computed from its own differential equation where the density is computed from P and T. Since the density figures in both the differential equations for P and for T, the three quantities affect one another. Working with polytropes, although physically less correct, has the advantage of simplifying the model considerably because the complicated equation for the temperature disappears from the method.

8.2.2 The Equation of State

This equation, linking the pressure, the density, the temperature, and the chemical composition, describes the gas properties of stellar matter. In normal terrestrial conditions, the equation is the perfect gas law. In the stellar interior, however, the temperature is so high that the corresponding radiation pressure represents a significant contribution to the total pressure. Thus, the total pressure P is the sum of the gas pressure P_G and the radiation pressure

P_R. The importance of the radiation pressure increases with the total mass of the star. The complete equation of state is

$$P = P_G + P_R = \frac{R\rho T}{\mu} + \frac{1}{3}aT^4 \tag{11}$$

where

T = the temperature in Kelvin (K)

R = the gas constant = 8.314×10^7(erg/mol/K)

μ = the mean molecular weight

a = the Stefan-Boltzmann constant = 7.56464×10^{-15}(erg/cm^3/K^4)

Since we concentrate on homogeneous stellar models, the mean molecular weight is constant throughout the star. The symbols X, Y, and Z are used to give the fractions by weight of hydrogen, helium, and all the other chemical elements. We will apply the galactic abundances normally used in professional models:

$$\begin{aligned} X &= 0.70 \text{ (hydrogen)} \\ Y &= 0.27 \text{ (helium)} \\ Z &= 0.03 \text{ (all other elements)} \end{aligned}$$

meaning that in 1 g of stellar matter there is 0.70 g of hydrogen, 0.27 g of helium, and 0.03 g of other elements. The mean molecular weight in the stellar interior is given by

$$\mu = \frac{2}{1 + 3X + 0.5Y} \tag{12}$$

or

$$\mu = \frac{2}{1 + 3(0.70) + 0.5(0.27)} = 0.618238. \tag{13}$$

It is convenient to express the gas pressure P_G and the radiation pressure P_R in units of the total pressure P by introducing a parameter β

$$P_G = \beta P; \;\; P_R = (1 - \beta)P. \tag{14}$$

The equation of state may then be written

$$P = \frac{R\rho T}{\mu\beta}. \tag{15}$$

For low mass stars, β is very close to 1. For a $10M_\odot$ star, it is about 0.95 in the core. By using the parameter β, the total pressure is written only by a number of multiplications and divisions. This allows the equation to be written in logarithmic form.

8.2.3 The Temperature

Since our model is based on polytropes, the temperature is computed as a function of the pressure and the density by means of the equation of state. The temperature itself has no effect on the pressure or density. In fact, the equation of state (11) for a given density and pressure is a polynomial of the fourth degree in the unknown temperature T. This polynomial is easily solved by iterations using the formula

$$T_{i+1} = \frac{\mu}{R\rho}\left(P - \frac{1}{3}aT_i^4\right) \tag{16a}$$

starting from the initial value

$$T_0 = \frac{\mu P}{R\rho}. \tag{16b}$$

This iterative procedure converges rapidly since the magnitude of the radiation pressure is small compared to the total pressure. Even in the core of a $15M_\odot$ star, the error is less than 0.01% after 10 iterations. In the author's program, a fixed number of ten iterations is computed each time T has to be computed from the density and the pressure. A more elegant way of calculating T is to work with a variable number of iterations controlled by a convergence criterion, for instance, the relative difference of two subsequent iterations being less than 0.001%.

- *Example:* $P = 2.5215 \times 10^{16}$ and $\rho = 5.3365$. Find the temperature.

$$\begin{aligned}
T_0 &= 3.513566 \times 10^7 \\
T_1 &= 2.978079 \times 10^7 \\
T_2 &= 3.237190 \times 10^7 \\
T_3 &= 3.127707 \times 10^7 \\
&\vdots \\
T_9 &= 3.161963 \times 10^7 \\
T_{10} &= 3.162343 \times 10^7
\end{aligned}$$

8.2.4 The Luminosity

Even a small star like our Sun produces an enormous amount of energy: 3.83×10^{33} erg/sec. For massive stars, the luminosity (or energy production) is much higher: it is more or less proportional to the third power of the mass of the star. The only physical processes capable of producing so much energy for such a long time are nuclear reactions. In the cores of the stars we are dealing with here, there are two nuclear reaction chains to be included that convert hydrogen into helium.

The temperatures in the stellar interior are high enough to ionize all elements completely. This means that of a hydrogen atom, (normally composed of a proton and an electron) only the proton is left, while the electron moves freely in the hot plasma of the stellar interior. In the same way, a helium nucleus consists of two protons and two neutrons without the normal pair of electrons.

A direct consequence of this situation is that all free particles in the interior are electrically charged; the nuclei (naked cores) have a positive charge equal to the number of protons, and the free electrons have a negative charge of one unit. It is a well-known fact that equal charges reject each other. Two hydrogen nuclei will, therefore, normally repulse before they may collide. However, the particles move at very high speed and are not always able to avoid each other in time since the stellar temperatures are so high. The result of such a collision of two hydrogen nuclei (i.e., two protons) is the formation of a new particle and some minor products as a small amount of energy is released. Such a process is called *nuclear fusion.* The two relevant nuclear fusion reactions in our case are discussed below.

1) The proton-proton cycle (pp-cycle)

$$\mathrm{H}_1^+ + \mathrm{H}_1^+ \longrightarrow \mathrm{D}_2^+ + \mathrm{e}^+ + \nu$$

$$\mathrm{D}_2^+ + \mathrm{H}_1^+ \longrightarrow \mathrm{He}_3^{++} + \gamma$$

$$\mathrm{He}_3^+ + \mathrm{He}_3^+ \longrightarrow \mathrm{He}_4^{++} + \mathrm{H}_1^+ + \mathrm{H}_1^+$$

In the first step, two hydrogen nuclei (the subscript referring to the mass, the superscript to the charge) form a deuteron consisting of one proton and one neutron, a positron and a neutrino. This deuteron meets in the second step a third hydrogen nucleus and forms a helium(3) nucleus and a small amount of gamma radiation. Finally, two of such helium nuclei form a normal helium(4) nuclei and two protons. Globally, six protons are needed to form one helium nucleus and two protons. In other words four hydrogen nuclei can produce one helium nucleus. The total mass of all the input elements is a little bit higher than the total mass of the resulting elements after fusion. The mass deficit is converted into energy according to Einstein's equation

$$E = mc^2. \tag{17}$$

This pp-cycle is important in the stellar core of low mass stars such as our Sun. As a consequence of nuclear reactions, the mass of the Sun slowly decreases. This decrease amounts to about 0.3% over the total duration of the hydrogen burning stage.

2) The CNO-cycle

This reaction uses carbon as a catalyst to capture four protons going over nitrogen and oxygen to finally break up into a new carbon nucleus of the same type as the original one and one helium ion. Therefore, the net production is once again four hydrogen nuclei into one helium nucleus. The mass deficit is converted into energy. This second reaction becomes more and more important with increasing mass.

Since both reactions occur only at sufficiently high temperatures, the energy production of 1 g of stellar matter is a function of the temperature, the density, and the chemical composition. The energy production E may be computed from the following formulae:

$$t = \left(\frac{T}{10^9}\right)^{1/3} \tag{18a}$$

$$P1 = 1 + 0.133t + 1.09t^2 + 0.938t^3 \tag{18b}$$

$$P2 = 1 + 0.027t^1 - 0.788t^2 - 0.149t^3 + 0.261t^4 + 0.127t^5 \tag{18c}$$

$$E_{\mathrm{pp}} = 2.376 \times 10^4 t^{-2} P1 \exp\left(\frac{-3.38}{t}\right) \tag{18d}$$

$$E_{\mathrm{cno}} = 8.666510^{25} t^{-2} P2 \exp\left(\frac{-15.228}{t} - \frac{t^6}{9.5481}\right) \tag{18e}$$

and finally

$$E = \rho(X^2 E_{\mathrm{pp}} + 0.02 X E_{\mathrm{cno}}) \tag{18f}$$

where X is the hydrogen abundance (0.70). These expressions are taken from the stellar evolution program of the Free University of Brussels. They were based on the work of W.A. Fowler and collaborators during the early seventies (see e.g. *Ann. Rev. Astron. Astrophys.*, Vol. 13, 1975, for a complete review on their results) and were slightly modified for our simple model. With E being the energy production of 1 g, the energy increase in a shell with size dr at distance r from the stellar center is

$$dL_r = 4\pi\rho E r^2 dr. \tag{19}$$

The total energy production of the star (the luminosity) is the sum of the contributions dL_r of all the subsequent shells throughout the star. This formula completes the set of equations describing our simple stellar model. The next problems to be solved are of a more practical nature:

- What is the polytrope index in the core?
- At what level in the star should we switch from a convective polytrope to a radiative one?
- How do we fit the two polytropes?
- Where is the surface of the star?

8.3 Boundaries and Polytrope Fitting Procedure

8.3.1 The Polytrope Index in the Core

In the convective core of a massive star, the temperature gradient is proportional to the pressure gradient. The derivation of the equation for the T-gradient may be found in most textbooks on stellar structure:

$$\frac{dT}{dr} = f(\beta)\left(\frac{T}{P}\right)\left(\frac{dP}{dr}\right) \tag{20a}$$

where

$$f(\beta) = \frac{8 - 6\beta}{32 - 24\beta - 3\beta^2}. \tag{20b}$$

In the absence of radiation pressure (i.e., $\beta = 1$), $f(\beta)$ becomes 0.4.

We will now derive another form of the equation for dT/dr in order to find a value for the polytrope index of the convective core. We'll assume that b is constant throughout the core. This assumption is sufficiently justified by the results of professional models. Taking the logarithms of both sides of the equation of state (15), we find

$$\log P = \log R + \log\rho + \log T - \log\mu - log\beta.$$

Then compute the derivatives as follows:

$$\left(\frac{1}{P}\right)\left(\frac{dP}{dr}\right) = \left(\frac{1}{T}\right)\left(\frac{dT}{dr}\right) + \left(\frac{1}{\rho}\right)\left(\frac{d\rho}{dr}\right). \tag{21}$$

When we do the same in relation (3) (logarithms and derivatives) we find

$$\left(\frac{1}{P}\right)\left(\frac{dP}{dr}\right) = \left(\frac{n+1}{n}\right)\left(\frac{1}{\rho}\right)\left(\frac{d\rho}{dr}\right). \tag{22}$$

Combining (21) and (22) finally gives

$$\frac{dT}{dr} = \left(\frac{1}{n+1}\right)\left(\frac{T}{P}\right)\left(\frac{dP}{dr}\right). \tag{23}$$

Identification of the corresponding terms in (23) and (20) yields

$$n = \frac{1 - f(\beta)}{f(\beta)}. \tag{24}$$

In this way the polytrope index of the convective core is evaluated as a function of β. If $\beta = 1$, then $n = 1.5$ as it should for a convective index in the absence of radiation pressure. In our application, β will vary from 1 to 0.9, making n vary from 1.5 to about 2 for a $15M_\odot$ star. We will work with a constant value of β for the whole core. This value will be taken so that the average effect of the radiation pressure in the convective core is 2/3 of its central value (see 25d). This means that the polytrope index n in the core is also constant. In practice, first compute the central gas and radiation pressure from the central density and the central temperature:

$$P_{G,C} = \frac{R\rho_C T_C}{\mu} \tag{25a}$$

$$P_{R,C} = \frac{1}{3} a T_C^4 \tag{25b}$$

then in the center

$$\beta_c = \frac{P_{G,C}}{P_{G,C} + P_{R,C}} \tag{25c}$$

and the average value of β throughout the core is

$$\beta = 1 - \frac{2}{3}(1 - \beta_c) \tag{25d}$$

and finally the polytrope index n from (24). This completes the method to derive the polytrope index in the core.

8.3.2 The Position of the Boundary of the Convective Core

Since our model is composed of two polytropes, at some level in the star we have to switch from the inner convective polytrope to the outer radiative one with index $n = 3$. This level marks the boundary of the convective core. Its position is found by identifying two expressions for dT/dr. In the previous subsection 3.1, we already mentioned the form of the temperature gradient in a convective medium. In a radiative medium, it takes a completely different form. We refer to textbooks for a proof of dT/dr in that case. In the radiative zone

$$\frac{dT}{dr} = -\left(\frac{3\rho\kappa}{4acT^3}\right)\left(\frac{L_r}{4\pi r^2}\right) \tag{26}$$

where

$\kappa =$ the absorption coefficient. For the sake of simplicity we will assume that electron scattering is the only contributing physical process. In that case

$$\kappa = 0.2(1 + X) = 0.34$$

since the hydrogen abundance X is 0.70. This coefficient is independent of density and temperature.

$a =$ the Stefan-Boltzmann constant (see 2.1).

$c =$ the speed of light $= 3 \times 10^{10}$cm/s.

At the boundary of the core where both polytropes meet, the expressions for dT/dr should be equal. Identifying (20a) = (26), replacing all constants by their value and rearranging the terms, one finds

$$1.339944 \times 10^9 \frac{PL_r}{M_r T^4 f(\beta)} = 1. \tag{27}$$

This formula should be programmed in a special way. Since P and L_r are both very large numbers, the numerator is of the order 10^{58}, which is much too large for most computers, hence causing an overflow. This problem is easily solved by mixing the divisions and multiplications to rearrange the various terms so that we never exceed the upper limit of about 10^{38} during the calculation. A practical way to do this, taking into account the magnitudes of the terms of (27), is given at the end of section 5.3.

In the convective zone, the left-hand side of this expression is always larger than one; in the radiative zone it is smaller. Therefore, it is sufficient to compute (27) at every level in the star during the iteration procedure. As soon as the result is smaller than one, we have passed the boundary of the core.

To simplify the program we will define the layer at which we fit the radiative polytrope as the first layer where (27) is smaller than one. We'll label this first layer b. In this way, the size of the core is slightly overestimated, but the error is small and has only a small effect on the structure of the star. The exact size of the convective core may be computed by interpolation between the last layer, where (27) is larger than one (layer $b-1$), and the first layer, where it is smaller (the boundary layer b):

$$f = \frac{1 - (27 \text{ at } b-1)}{(27 \text{ at } b) - (27 \text{ at } b-1)}.$$

Then the mass of the convective core is

$$MCC = M_r(b-1) + f\big(M_r(b) - M_r(b-1)\big).$$

The next step deals with switching from the inner to the outer polytrope.

8.3.3 Fitting the Two Polytropes

When we arrive at the boundary layer, the values of all the physical quantities at that layer are already computed since we needed them to evaluate the boundary criterion. Let us label the values at the boundary with subscript b. Thus, at the boundary we have the following equivalents:

P_b = the pressure

r_b = the radius

ρ_b = the density

$M_{r,b}$ = the mass inside r_b

With these values we will now compute the initial values and conditions of the outer polytrope. Since we are no longer at the center of the star, it is no longer possible to start the outer polytrope iteration from the initial values $x = 0$, $F = 1$, and $H = 0$. Also, the two parameters of the outer polytrope P_c and ρ_c are no longer the real central pressure and central density. Therefore, we have to choose five initial parameters to start the iteration of the radiative polytrope:

$$x, \;\; F, \;\; H, \;\; P_c, \;\text{and}\;\; \rho_c.$$

These five parameters may not be chosen independently from each other. It is obvious that the stellar structure should be continuous at the boundary. At the boundary, therefore, the values of the pressure, the density, the distance from the stellar center, and the mass inside that radius obtained from the inner polytrope should be the same as their values from the outer polytrope at the boundary. As a result the five parameters should not violate the expressions (6,7,8,9) evaluated at the boundary level:

$$P_b = P_c F^{n+1} \tag{28}$$

$$\rho_b = \rho_c F^n \tag{29}$$

$$M_{r,b} = -4\pi r_n^3 \rho_c x^2 H \tag{30}$$

$$r_n = \sqrt{\frac{P_c}{\pi G \rho_c^2}} \tag{31a}$$

$$r_b = r_n x. \tag{31b}$$

These five equations, with the five parameters at the right-hand side as unknowns, must be satisfied to have a smooth fitting of the two polytropes. Since we have only four equations and five parameters, one out of the five may be chosen freely. The simplest parameter to choose freely in our model is F. Then P_c is immediately obtained from (28), ρ_c from (29), r_n from (31a), x from (31b), and H from (30). Since the other physical quantities such as the luminosity, energy production, and temperature are functions of the basic polytrope variables, their continuity is also guaranteed. In section 8.4 we present a formula to select the right value of the parameter F as a function of the total mass of the star.

8.3.4 Stopping the Iterations at the Stellar Surface

Since the polytrope index used is small enough, the function $F(x)$ that controls the pressure and density crosses the X-axis at some point x_S. This means that the pressure and density become zero so that the surface of the star is reached. The zero point x_S will normally be located between two subsequent mesh points of the iteration procedure. Since we apply a midpoint method, each new iteration requires the evaluation of $F_{i+1/2}$ and F_{i+1}. As soon as one of these two becomes negative, the surface of the star has been crossed and the iteration has to be stopped immediately. If $F_{i+1/2}$ turns out to be negative, do not calculate $H_{i+1/2}$ anymore, nor F_{i+1} or H_{i+1}; if it is F_{i+1} that has become negative, do not calculate H_{i+1}. Use the last layer where F_i was positive to compute the location of the surface of the star.

The position of the stellar surface may be located with sufficient accuracy from

$$x_S = x_L - \frac{F_L}{H_L} \tag{32}$$

where the subscript L refers to the last layer x_i where F_i was positive. For obvious reasons F_S is zero, but the value of H at the surface is still slightly negative and may be taken:

$$H_S = H_L. \tag{33}$$

The pressure, density, and temperature are zero at the surface, as they depend directly on F. Since the energy production in the radiative zone is zero, the luminosity is constant there. The surface luminosity is, therefore, equal to that constant value. The only two variables without a trivial value at the surface are the mass and the radius. The surface value of the mass (i.e., the total mass of the star) is obtained as the sum of the mass inside x_L obtained from the normal iterations, added to the small amount of mass between x_L and x_S. Taking half the density at x_L for the average density of this shell, the mass of the shell between x_L and x_S is

$$M = 2\pi r^2 \rho dr.$$

Writing this as a function of x and evaluating all quantities halfway between x_L and x_S, we finally find for the total mass of the star

$$M = M_L + 0.5\pi r_n^3 \rho_L (x_L + x_S)^2 (x_S - x_L) \tag{34}$$

and for the total radius of the star

$$R = r_L + r_n(x_S - x_L). \tag{35}$$

The total mass is obtained in grams, and the radius is obtained in centimeters when these expressions are used. Divide by 2×10^{33} and 6.96×10^{10}, respectively, to have results in solar units. All luminosities are in centimeter-gram-second units and should be divided by 3.98×10^{33} to express them in solar units.

The last important property of the star is the effective temperature. It is obtained from the total radius R and luminosity L from the relation

$$L = 4\pi R^2 \sigma T_{\text{eff}}^4.$$

We need the logarithm of the effective temperature to plot the position of the model in the HR diagram. With the luminosity L and radius R expressed in solar units we find

$$\log T_{\text{eff}} = 3.7613 + 0.25 \log L - 0.5 \log R. \tag{36}$$

The effective temperature is in Kelvin. Then to plot the model in the Hertzsprung-Russell diagram, $\log T_{\text{eff}}$ and $\log L$ are used. We refer to the last section for a number of applications. A definition of the effective temperature is described in the chapter on stellar atmospheres.

8.4 The Choice of the Free Parameters

Our models are characterized by two initial conditions (the central density and pressure) and one additional free parameter F_{fit}, the value of the Lane-Emden function $F(x)$ of the outer polytrope at the fitmass. Since it is usual to work with the central temperature instead of the pressure, we may replace P_C with T_C. With the three free choices (ρ_C, T_C, and F_{fit}), a large number of models may be computed. Important observational facts, however, limit this number considerably. The models are subject to two constraints, namely, the observed position of homogeneous stars in the HR diagram and the observed mass-luminosity relation. The effect of these constraints will be taken into account as set forth below.

Since the total energy production occurs in the convective core, the luminosity at the surface is not affected by the choice of F_{fit}. Hence, starting from a certain central density and pressure, the luminosity may be computed as the luminosity at the fitmass. This luminosity corresponds to a certain total mass of the star (mass-luminosity relation). If we now proceed in the radiative zone with an arbitrary choice of F_{fit}, we will find a total mass but not necessarily the mass suggested by the mass-luminosity relation. Only one choice of F_{fit} will satisfy the right mass. In this way the mass-luminosity relation limits the choice of F_{fit}. It is no longer a free parameter but depends on the central density and temperature.

Furthermore, various combinations of T_C and ρ_C may result in the same luminosity. For each of them F_{fit} may be taken to find the right mass at the surface. These models will all have a different radius, so that only one combination of T_C and ρ_C will give the right position in the HR diagram. In this way the choice of T_C and ρ_C is limited to only one couple giving the right position in the HR diagram.

These two constraints mean that in reality only one parameter may be chosen freely. One could take, for instance, T_C as the free parameter and compute the only "good" model. By considering various values of T_C, we then obtain a series of models as a one-parameter function of the central temperature. These models all have their own central density and their own value of F_{fit}. They also have a certain mass M. It is then possible to derive formulae giving the three initial parameters as functions of the mass. In this way we are able to use the total mass, the most important parameter of a star, as the only free parameter. In order to be able to compare our models with professional results, we will use interpolation formulae from professional programs for the central density and temperature. Only the value of F_{fit} will be chosen freely to reproduce more-or-less the main properties of the stars. Fortunately, our models depend only slightly on the choice of the parameter F_{fit}.

The fits, with the total mass M expressed in solar units, are as follows:

$$w = \log M$$

$$\log T_C = 7.23937 + 0.2724354\, w - 0.0401771\, w^2 \tag{37a}$$

$$\log \rho_C = 2.27899 - 1.658707\, w + 0.29329095\, w^2 \tag{37b}$$

$$\left.\begin{array}{ll} F_{\text{fit}} = 9 & \text{if } M < 4 \\ F_{\text{fit}} = 19.58794 - 17.58794\, w & \text{if } 4 < M < 10 \\ F_{\text{fit}} = 2 & \text{if } M > 10 \end{array}\right\} \tag{37c}$$

Do not forget that (37a,b) give the logarithms of the central temperature and density and should therefore be transformed into the normal values before application in the numerical method.

With these expressions, the three free parameters are easily obtained, based on the total mass, and produce models consistent with the observational constraints.

8.5 Numerical Method

We will discuss in this section the subsequent steps of the calculation of our stellar model. The most important part of the solution is the solution of the Lane-Emden equation for the structure of the polytropes. The step-size for the numerical iteration will be called dx, and a step-size $dx = 0.1$ will be used in the convective core of all our models. Outside the core in the radiative zone, the density decreases towards zero at the surface. There, use of a constant step-size is not recommended. We will start at the bottom of the radiative polytrope with $dx = 0.4$ and increase dx by 10% with every step. So, after every step $dx = 1.1dx$. This process yields models of about twenty layers for the low masses to more than thirty for masses up to $15M_{\odot}$.

8.5.1 Initial Conditions and Central Values

It is obvious that the calculation of T from P and ρ from (16a,b) should be placed in a subprogram. The same holds for E from T and ρ with (18a–f). The only free parameter, the total mass, may be selected between 2 and $15M_{\odot}$. With your choice of the total mass, the following quantities are computed:

- the central temperature from (37a)
- the central density from (37b)

- the central gas pressure from (25a) and the central radiation pressure from (25b)
- the total central pressure as their sum
- the value of β in the convective core from (25c,d)
- the value of $f(\beta)$ in the convective core from (20b)
- the convective core polytrope index from (24)
- the distance parameter of the convective core from (10)
- the initial conditions of the Lane-Emden equation $x = 0$; $F = 1$; $H = 0$
- the value of F_{fit} of the outer polytrope at the fitmass from (37c)
- the central nuclear energy production E_C from (18a–f) from T_C and ρ_C
- the central values of the mass M, the radius r, and the luminosity L are, of course, zero.

This completes the set of initial conditions. The next point is to compute the first step from $x = 0$ to $x = dx$.

8.5.2 The First Step

Since we deal with polytropes, the structure of the model is given by the solution of the Lane-Emden equation. This means that it is not possible to start with the normal iteration procedure since $x_0 = 0$. This would mean that we should have to divide by zero in formula (40c). The first step is, therefore, performed by using polynomial expansions of $F(x)$ and $H(x)$ around $x = 0$. These expansions are, in fact, Taylor series developed at $x = 0$. If we keep close enough to $x = 0$, as we do for the first step, these expansions are sufficiently accurate. In this case with $x = dx$

$$F_1 = 1 - \frac{1}{6}dx^2 + \frac{n}{120}dx^4 \tag{38a}$$

$$H_1 = -\frac{1}{3}dx + \frac{n}{30}dx^3 \tag{38b}$$

and

$$x_1 = dx \tag{38c}$$

n stands for the polytrope index of the core. The pressure, density, mass, and radius are then computed from (6,7,8,9). The temperature T_1 at the first level is computed by iteration of (16a), starting with (16b) as the initial value.

The luminosity at the first level is also computed by a Taylor series around zero. Thus

$$L_1 = \frac{4}{3}\pi \rho_C E_C dr^3 \tag{39}$$

in which

$$dr = r_n dx.$$

This completes the calculations of the first layer. From now on, the normal midpoint iterations will be used up to the boundary of the core.

8.5.3 The Convective Core Iteration

The main part of this section is the solution of the Lane-Emden equation which enables us to proceed from layer i to layer $i+1$. The formulae are suited for starting from $i = 1$. Let us proceed, beginning with the values of x, F, and H at layer i.

$$x_{i+1/2} = x_i + \frac{1}{2}dx \tag{40a}$$

$$F_{i+1/2} = F_i + \frac{1}{2}dx H_i \tag{40b}$$

$$H_{i+1/2} = H_i + \frac{1}{2}dx\left(-F_i^n - \frac{2}{x_i}H_i\right). \tag{40c}$$

Some physical quantities we also need to know halfway through the layer are

$$P_{i+1/2} = P_C F_{i+1/2}^{n+1}$$

$$\rho_{i+1/2} = \rho_C F_{i+1/2}^n.$$

With these two: $T_{i+1/2}$ with (16a,b) and $E_{i+1/2}$ with (18a–f).
Then compute:

$$x_{i+1} = x_i + dx$$

$$F_{i+1} = F_i + dx H_{i+1/2}$$

$$H_{i+1} = H_i + dx\left(-F_{i+1/2}^n - \frac{2}{x_{i+1/2}}H_{i+1/2}\right).$$

And for the physical variables:

$$P_{i+1} = P_C F_{i+1}^{n+1}$$

$$\rho_{i+1} = \rho_C F_{i+1}^n$$

M_{i+1} with (8) evaluated at x_{i+1} and H_{i+1}

$$r_{i+1} = r_n x_{i+1}$$

$$T_{i+1} \text{ with (16a, b) and } P_{i+1} \text{ and } \rho_{i+1}.$$

The luminosity at layer $i+1$ is obtained from equation (19) which becomes

$$L_{i+1} = L_i + 4\pi \rho_{i+1/2}\, E_{i+1/2}\, r_n^3\, x_{i+1/2}^2\, dx.$$

The fact that we use the density and the energy production at half the step is consistent with the use of the midpoint method:

$$E_{i+1} \text{ from (18a–f) with } T_{i+1} \text{ and } \rho_{i+1}.$$

This completes the computation of all the variables at level $i+1$. These formulae are used until the boundary of the convective core is reached. It is, therefore, necessary to compute at each new level the left hand side of (27). As soon as it is smaller than one, the midpoint iteration is stopped, and the fitting method for the outer radiative polytrope is performed. The last layer of the core is also the first of the radiative envelope. We will label this special layer with the index b. Eventually, the correct size of the convective core may be interpolated between layer b and $b-1$.

We mentioned in section 3.2 the need to rearrange the terms in equation (27) in order to avoid an overflow. Equation (27) may safely be computed in the following way:

$$\left(1.33994 \times 10^9\right) \times \left(\frac{P}{M_r}\right) \times \left(\frac{L_r}{T^4}\right) \times \left(\frac{1}{f(\beta)}\right).$$

Take turns in computing the numerators and denominators. When you try to complete the product of all the numerators, an overflow error will occur.

8.5.4 Fitting the Two Polytropes

From now on the polytrope index equals 3! The boundary values of the outer polytrope and the outer polytrope parameters P_C and ρ_C are obtained with the formulae of section 3.3. F equals F_{fit}, already computed at the start of our program with (37c). The new values of P_C and ρ_C are then found from (28) and (29), x from (31) and H from (30). With these five values we are sure that the outer polytrope fits smoothly to the core. Finally, we compute with these parameters the new value of r_n from (9) and of course with $n = 3$.

It is important to realize that if you want to use the same piece of program for the outer polytrope, the test of formula (27) must not be executed in the radiative zone since, in that case, you would always continue to fit new polytropes after every step in the radiative zone. This problem may easily be solved by adding a condition to (27):

Fit a radiative polytrope if (27) < 1 *and* if $n < 2.5$.

In the core only the second condition is satisfied. Once we have reached its boundary, both conditions are satisfied, and a new polytrope of index 3 is fitted. From that level on, only the first condition is satisfied so that we may be sure that the fitting procedure will not be executed any more. Note that the negation of the double condition is:

Do not fit a radiative polytrope if (27) > 1 *or* if $n > 2.5$.

If you program the formulae of 5.3 a second time for the radiative zone, the single test of (27) is sufficient, but your program will be longer. The flow chart presented on the following page uses this easy way.

8.5.5 The Radiative Zone

The formulae for this part are exactly the same as for the convective core since the parameters P_C, ρ_C, r_n, and n were updated in the fitting formulae. Also the starting values of x, F, and H were computed. The only change is that now we no longer compute the nuclear energy production (there isn't any) or any new value or the luminosity since it doesn't increase any more. An important change, however, is the step-size $dx = 1.1dx$ at every new step, starting with $dx = 0.4$ for the first step at the bottom of the radiative zone.

The surface of the star is reached when F becomes zero or negative. This may occur after the computation of a $F_{i+1/2}$ or a F_{i+1}. The test on F being negative therefore *must* be performed at both occasions in order to avoid a runtime error. As soon as one of these two is negative, the normal midpoint iteration is stopped without even calculating a new $H_{i+1/2}$ or a new H_{i+1}. Finally, the surface quantities are computed.

8.6 The Final Results

The exact position of the surface is found starting from the polytrope solution x_i, F_i, H_i for which F was positive. The zero point is found from (32) and the value of H at that point from (33). The total mass, total luminosity, and the effective temperature are obtained from (34,35,36).

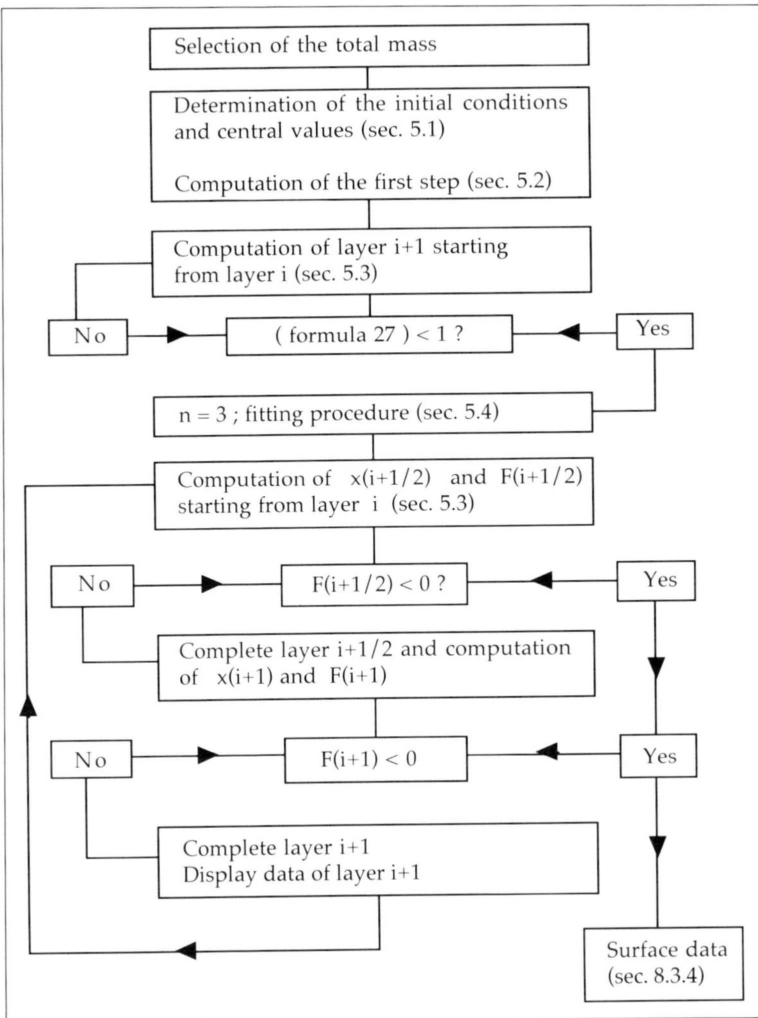

8.7 Applications

8.7.1 Model of a $10M_\odot$ Star

The initial parameters for a $10M_\odot$ star are as follows:

$$\log T_C = 7.47163$$
$$\log \rho_C = 0.913574$$
$$F_{\text{fit}} = 2$$

thus

$$\begin{aligned} \beta &= 0.9625768 \\ n(\text{core}) &= 1.7504546. \end{aligned}$$

Some other central values are

$$\begin{aligned} \log P_c &= 16.53895 \text{ (total gas pressure), and} \\ \log E &= 4.61177 \text{ (energy production).} \end{aligned}$$

At layer 16 when $x = 1.6$ for the convective polytrope, the condition to fit a radiative polytrope is satisfied. Based on the value of F_{fit} and the data of layer 16, we find the following as new parameters for the outer polytrope:

$$\begin{aligned} P_C &= 6.7110721 \times 10^{14} \\ \rho_C &= 0.486585 \\ x &= 0.5655283 \\ H &= -2.069755 \\ r_n &= 1.1627904 \times 10^{11} \end{aligned}$$

We then proceed in the radiative zone with $n = 3$. The results on the following pages clearly show that the physical quantities (pressure, density, distance, mass...) vary smoothly over the fitlayer, although the polytrope variables x, F, and H do not. The last value of x for which F is positive is $x_L = 2.3895$.

Following are the final results:

$$\begin{aligned} \text{total mass} &= 10.33\ M_\odot \\ \text{radius} &= 4.08\ R_0 \\ \log T_{\text{eff}} &= 4.40\ K \\ \log L/L_0 &= 3.80 \end{aligned}$$

Since our program works with centimeter-gram-second units, we have transformed these results in solar units. (See appendix for all physical constants.) The Tables 8-1 and 8-2 show the results for the $10 M_\odot$ model.

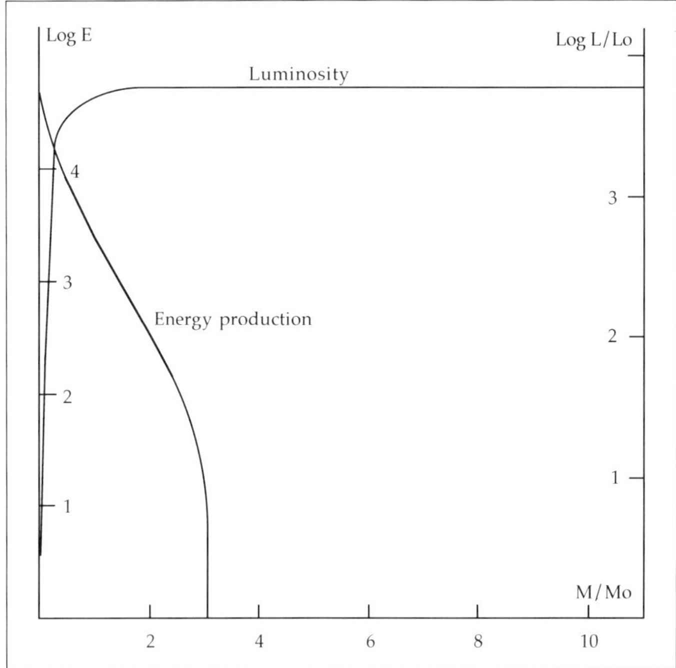

Fig. 8.1: *The variation of the energy production and the luminosity inside a* $10M_\odot$ *as functions of the mass. The complete energy production is concentrated in a small sphere around the center of the star. The total luminosity of the star is reached at its border and stays constant outside that sphere since there is no additional energy production.*

8.7.2 Results for Other Masses

We will now compare our results for other masses with the data from professional programs. These data were computed with the program for stellar evolution of the Astrophysical Institute of the Free University of Brussels, Belgium. Table 8-3 lists the relevant parameters for a number of masses between 2 and $15M_\odot$. At the left side are the professional results; at the right side, the data from our composite polytrope models. Our composite polytropes were computed with the same initial conditions (central density and temperature) as the professional models.

The masses of our models are a little bit too large: 10% for the low masses to about 2% for the high masses. The radii are quite well reproduced except for the lowest mass where the deviation is 17%. All our models are overluminous and too hot, but these deviations are attenuated by the fact that the masses are also higher. Once again, the deviations are the largest

Table 8-1

i	M	$\log P$	$\log T$	$\log \rho$	r
0	0.00000	16.5389	7.4716	0.9136	0.00000
1	0.00119	16.5370	7.4709	0.9123	0.05905
2	0.00947	16.5310	7.4689	0.9085	0.11810
3	0.03168	16.5210	7.4655	0.9022	0.17715
4	0.07417	16.5071	7.4607	0.8933	0.23620
5	0.14261	16.4891	7.4546	0.8819	0.29525
6	0.24177	16.4673	7.4471	0.8680	0.35431
7	0.37539	16.4414	7.4382	0.8515	0.41336
8	0.54610	16.4116	7.4280	0.8325	0.47241
9	0.75531	16.3779	7.4164	0.8111	0.53146
10	1.00326	16.3402	7.4034	0.7871	0.59051
11	1.28900	16.2986	7.3890	0.7606	0.64956
12	1.61049	16.2530	7.3733	0.7316	0.70861
13	1.96473	16.2035	7.3561	0.7001	0.76766
14	2.34782	16.1500	7.3376	0.6660	0.82671
15	2.75520	16.0925	7.3176	0.6294	0.88576
16	3.18179	16.0309	7.2961	0.5902	0.94481
17	3.67581	15.9570	7.2776	0.5348	1.01164
18	4.21956	15.8716	7.2563	0.4708	1.08515
19	4.80764	15.7734	7.2317	0.3971	1.16601
20	5.43092	15.6607	7.2036	0.3126	1.25496
21	6.07643	15.5321	7.1714	0.2161	1.35280
22	6.72768	15.3856	7.1348	0.1063	1.46043
23	7.36566	15.2194	7.0933	0.0184	1.57881
24	7.97033	15.0313	7.0462	0.1595	1.70904
25	8.52256	14.8184	6.9930	0.3191	1.85229
26	9.00618	14.5775	6.9328	0.4998	2.00987
27	9.40987	14.3040	6.8644	0.7050	2.18320
28	9.72849	13.9909	6.7861	0.9398	2.37386
29	9.96373	13.6277	6.6953	1.2121	2.58360
30	0.12384	13.1966	6.5876	1.5355	2.81430
31	0.22262	12.6643	6.4545	1.9347	3.06808
32	0.27767	11.9588	6.2781	2.4639	3.34723
33	0.30775	10.8751	6.0072	3.2766	3.65430
34	0.32916	8.0784	5.3080	5.3742	3.99207

for the lower masses. We feel that the observed mass-luminosity relation and the place of the Zero Age Main Sequence are sufficiently well reproduced by our models, although they were constructed using very simple physical laws compared with the physics involved in professional computer codes. It would be possible to reproduce the observations more accurately by changing the initial conditions for the central density and temperature. We decided to take the initial conditions of professional programs to be sure to start from correct values and to be able to compare our models. We did, however, take the liberty of selecting the value of $F_{\rm fit}$ to obtain more-or-less the right global characteristics of the stars.

It is also possible to compare our models to professional models with the same mass instead of the same central temperature and density. The differences in the effective temperature and the luminosity between our models and the professional models then nearly vanish for most of the masses considered, since a slight increase of the mass of the professional models also means a slight increase of their temperature and of their luminosity. For instance, a professional model with a mass of 10.33M has an effective temperature of Log $T_{\rm eff} = 4.40$ (the same as with our model) and a luminosity of Log $L = 3.77$ (3.80 with our model).

Table 8-2

i	M	$\log E$	$\log L$	x	f	h
0	0.00000	4.6118	0.0000	0.0000	1.0000	0.0000
1	0.00119	4.5998	1.4057	0.1000	0.9983	−0.0333
2	0.00947	4.5639	2.2721	0.2000	0.9933	−0.0662
3	0.03168	4.5040	2.7669	0.3000	0.9851	−0.0985
4	0.07417	4.4198	3.0910	0.4000	0.9737	−0.1297
5	0.14261	4.3111	3.3163	0.5000	0.9592	−0.1596
6	0.24177	4.1776	3.4760	0.6000	0.9418	−0.1879
7	0.37539	4.0189	3.5888	0.7000	0.9216	−0.2143
8	0.54610	3.8346	3.6668	0.8000	0.8989	−0.2387
9	0.75531	3.6240	3.7191	0.9000	0.8739	−0.2608
10	1.00326	3.3866	3.7526	1.0000	0.8467	−0.2806
11	1.28900	3.1216	3.7730	1.1000	0.8177	−0.2980
12	1.61049	2.8285	3.7846	1.2000	0.7871	−0.3128
13	1.96473	2.5069	3.7908	1.3000	0.7552	−0.3252
14	2.34782	2.1568	3.7939	1.4000	0.7221	−0.3351
15	2.75520	1.7798	3.7952	1.5000	0.6881	−0.3425
16	3.18179	1.3811	3.7958	1.6000	0.6536	−0.3477
17	3.67581	1.3811	3.7958	0.6055	1.9167	−2.0856
18	4.21956	1.3811	3.7958	0.6495	1.8247	−2.0808
19	4.80764	1.3811	3.7958	0.6979	1.7244	−2.0534
20	5.43092	1.3811	3.7958	0.7512	1.6162	−2.0024
21	6.07643	1.3811	3.7958	0.8097	1.5008	−1.9281
22	6.72768	1.3811	3.7958	0.8742	1.3795	−1.8317
23	7.36566	1.3811	3.7958	0.9450	1.2536	−1.7159
24	7.97033	1.3811	3.7958	1.0230	1.1249	−1.5846
25	8.52256	1.3811	3.7958	1.1087	0.9952	−1.4424
26	9.00618	1.3811	3.7958	1.2030	0.8663	−1.2946
27	9.40987	1.3811	3.7958	1.3068	0.7401	−1.1464
28	9.72849	1.3811	3.7958	1.4209	0.6181	−1.0025
29	9.96373	1.3811	3.7958	1.5464	0.5015	−0.8668
30	10.12384	1.3811	3.7958	1.6845	0.3913	−0.7422
31	10.22262	1.3811	3.7958	1.8364	0.2880	−0.6306
32	10.27767	1.3811	3.7958	2.0035	0.1919	−0.5327
33	10.30775	1.3811	3.7958	2.1873	0.1028	−0.4482
34	10.32916	1.3811	3.7958	2.3895	0.0206	−0.3764

Table 8-3

M	R	$\log T_e$	$\log L$	M	R	$\log T_e$	$\log L$
2	1.62	3.96	1.20	2.18	1.38	4.02	1.30
3	2.04	4.08	1.88	3.33	1.92	4.11	1.97
5	2.72	4.22	2.71	5.43	2.72	4.24	2.78
8	3.61	4.34	3.42	8.37	3.55	4.36	3.48
10	4.09	4.39	3.73	10.33	4.08	4.40	3.80
12	4.35	4.43	3.98	12.28	4.54	4.44	4.05
15	5.11	4.47	4.26	15.36	5.20	4.49	4.35

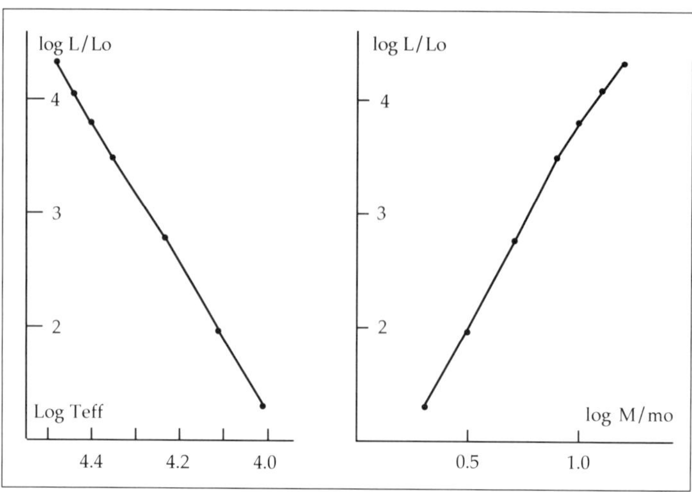

Fig. 8.2: *Computed mass-luminosity relation (right) and place in the HR-diagram (left).*

8.8 The Program Listing

The following sample program is written in MicroSoft QuickBasic 4.5:

```
DECLARE FUNCTION lg10 (x)
DECLARE FUNCTION temp (mu, p, d)
DECLARE FUNCTION e (d, t)
DECLARE SUB poly (n, x, f, h, pc, dc, rn, p, d, mr, r)

COMMON SHARED pi, g, a, rgas, xx, yy, zz, mu, m0, r0, l0

' declare a number of constants used in the program
pi = 3.1415926536#
g = 6.673E-08
a = 7.56464E-15
rgas = 8.314E+07
xx = .7
yy = .27
zz = .03
mu = .618238
m0 = 2E+33
r0 = 6.96E+10
l0 = 3.83E+33


CLS
PRINT " "
PRINT "Astrophysics With a PC : STELLAR MODEL"
PRINT "-----------------------------------------"
PRINT " "
PRINT "----------- Minimal solution program -------"
PRINT " "
INPUT "Input parameter : Approximation; of; total; mass(2 - 15) : ", mtot
```

```
' show heading of main table
PRINT " i  Mr/Mo  log(p)  log(T)  log(d)  r/ro   log(E)_
  log(L)   x      f       h"

w = lg10(mtot)

' central values of temperature and density
tc = 7.23937 + .2724354# * w - .0401771# * w * w
tc = 10 ^ tc
dc = 2.27899 - 1.658707 * w + .29329095# * w * w
dc = 10 ^ dc

' compute the value of the fitmass :
IF mtot < 4 THEN
    ffit = 9
ELSE
    IF m < 10 THEN
        ffit = 19.58794 - 17.58794 * w
    ELSE
        ffit = 2
    END IF
END IF

' compute other central quantities
pgc = rgas * dc * tc / mu
prc = 1 / 3 * a * tc ^ 4
ptc = pgc + prc
betac = pgc / ptc
beta = 1 - 2 / 3 * (1 - betac)
fb = (8 - 6 * beta) / (32 - 24 * beta - 3 * beta ^ 2)
n = (1 - fb) / fb
rn = SQR((n + 1) * ptc / 4 / pi / g / dc / dc)
eec = e(dc, tc)
i% = 0
x = 0
f = 1
h = 0
m = 0
r = 0
l = 10
CALL poly(n, x, f, h, ptc, dc, rn, p, d, m, r)

' print first line of table (central values)
PRINT USING "## ##.### ###.### ###.### ###.#### ##.### ###.### ###.### ##.###_
 ##.### ###.###"; i%; m / m0; lg10(p); lg10(tc); lg10(d); r / r0;_
 lg10(eec); lg10(l / 10); x; f; h

' compute and show first step
i% = 1
l = 0
dx = .1
x = dx
f = 1 - 1 / 6 * dx ^ 2 + n / 120 * dx ^ 4
h = -1 / 3 * dx + n / 30 * dx ^ 3
CALL poly(n, x, f, h, ptc, dc, rn, p, d, m, r)
t = temp(mu, p, d)
ee = e(d, t)
dr = dx * rn
l = 4 / 3 * pi * dc * eec * dr ^ 3
PRINT USING "## ##.### ###.### ###.### ###.#### ##.### ###.### ###.### ##.###_
```

```
 ##.### ###.###"; i%; m / m0; lg10(p); lg10(t); lg10(d); r / r0;_
 lg10(ee); lg10(l / 10); x; f; h


' Start of main cycle

i% = 2
FOR zone% = 1 TO 2        'zone% = 1 during convective region
                          '       = 2 during radiative region
    stp% = 0
    DO
        flast = f   'Save previous value of polytrope variable in flast
        x12 = x + .5 * dx
        f12 = f + .5 * dx * h


'       check whether f12 is still positive
        IF f12 > 0
        THEN
            h12 = h + .5 * dx * (-f ^ n - 2 * h / x)
            p12 = ptc * f12 ^ (n + 1)
            d12 = dc * f12 ^ n
            t12 = temp(mu, p12, d12)


'           compute energy production when in convective zone
            IF zone% = 1 THEN ee12 = e(d12, t12)
            x = x + dx
            f = f + dx * h12


'           check whether f is still positive
            IF f > 0
            THEN
                h = h + dx * (-f12 ^ n - 2 * h12 / x12)
                p = ptc * f ^ (n + 1)
                d = dc * f ^ n
                m = -4 * pi * dc * rn ^ 3 * x ^ 2 * h
                r = rn * x
                t = temp(mu, p, d)


'               compute new value of luminosity when in convective zone
                IF zone% = 1 THEN
                    l = l + 4 * pi * d12 * dx * ee12 * rn ^ 3 * x12 ^ 2
                END IF


'               compute energy production for new state when in convective
'               zone. In radiative zone, ee is put to 1, so that log(ee)_
 becomes 0
                IF zone% = 1
                THEN
                    ee = e(d, t)
                ELSE
                    ee = 1
                END IF


'               show results of layer  i+1  on screen
                PRINT USING "## ##.### ###.### ###.### ###.#### ##.###_
 ###.### ###.### ##.### ##.### ###.###"; i%; m / m0; lg10(p);_
 lg10(t); lg10(d); r / r0; lg10(ee); lg10(l / 10); x; f; h
                i% = i% + 1

            ELSE
                surface% = 1   'surface has been reached
            END IF
        ELSE
```

```
            surface% = 1        'surface has been reached
        END IF

'       check if in convective zone
        IF zone% = 1
        THEN
            test = 1.339944E+09 * p / m * l / t ^ 4 / fb

'           check if boundary of convective zone is reached
            IF test < 1
            THEN
                PRINT "Boundary of convective core is reached. Press any key"
                DO
                LOOP WHILE INKEY$ = ""

'               compute fitting parameters
                f = ffit
                ptc = p / ffit ^ 4
                dc = d / ffit ^ 3
                rn = SQR(ptc / pi / g / dc ^ 2)
                x = r / rn
                h = -m / 4 / pi / dc / rn ^ 3 / x ^ 2
                n = 3
                dx = .04
                stp% = 1
                surface% = 0

'               show fitting parameters
                PRINT " "
                PRINT "---------- Fitting parameters ----------"
                PRINT "New centr.press.    : "; ptc
                PRINT "New centr.densiity  : "; dc
                PRINT "xfit                : "; x
                PRINT "ffit                : "; ffit
                PRINT "hfit                : "; h
                PRINT "new rn              : "; rn
                PRINT " "

                PRINT "Press any key to continue"
                DO
                LOOP WHILE INKEY$ = ""
                PRINT " "
            END IF

'       next ELSE is entered when in radiative zone
        ELSE
            dx = 1.1 * dx

'           check if surface is reached
            IF surface% = 1 THEN

'               compute exact location of surface and surface data
                stp% = 1
                xs = x - flast / h
                mt = m + .5 * pi * d * rn ^ 3 * (x + xs) ^ 2
                rad = r + rn * (xs - x)
                logteff = 3.7613 + .25 * lg10(l / 10) - .5 * lg10(rad / r0)

                PRINT "Press any key to continue"
                DO
                LOOP WHILE INKEY$ = ""
```

```
                PRINT " "

'              print surface data
                PRINT "----------------------- surface data_
 ----------------------"
                PRINT USING "Mass (in Mo)           : ####.##"; mt / m0
                PRINT USING "radius(Ro)             : ####.##"; rad / r0
                PRINT USING "Luminosity (log(L/Lo): ####.##"; lg10(l / 10)
                PRINT USING "Effect.temp.(log)     : ####.##"; logteff
                PRINT " "
                PRINT "Press any key to stop this program"
                DO
                LOOP WHILE INKEY$ = ""
            END IF
        END IF

        LOOP UNTIL stp% = 1

    NEXT zone%
END


FUNCTION e (d, t)
'
' computes the energy production for given density d and temperature t
'
     tt = EXP(1 / 3 * LOG(t / 1E+09))
     p1 = 1 + tt * (.133 + tt * (1.09 + tt * .938))
     p2 = 1 + tt * (.027 + tt * (-.788 + tt * (-.149 + tt_
 * (.261 + tt * .127))))
     e1 = 23760! / tt ^ 2 * p1 * EXP(-3.38 / tt)
     e2 = 8.6665E+25 / tt ^ 2 * p2 * EXP(-15.228 / tt - tt ^ 6 / 9.5481)
     e = d * (xx ^ 2 * e1 + .02 * xx * e2)
END FUNCTION


FUNCTION lg10 (x)
'
' computes the base 10 logarithm of x using the natural logarithm LOG
'
     lg10 = LOG(x) / LOG(10!)
END FUNCTION


SUB poly (n, x, f, h, pc, dc, rn, p, d, mr, r)
'
' computes pressure, density, mass and distance to the centre starting
' from the polytrope results (x,f and h) and parameters n, pc and dc
'
     p = pc * f ^ (n + 1)
     d = dc * f ^ n
     mr = -4 * pi * dc * rn ^ 3 * x ^ 2 * h
     r = rn * x
END SUB


FUNCTION temp (mu, p, d)
'
' solves the equation of state to compute the temperature from the
' pressure, density and mean molecular weight
'
     tt = mu * p / rgas / d
     FOR i% = 1 TO 10
         tt = mu / rgas / d * (p - 1 / 3 * a * tt ^ 4)
```

```
    NEXT i%
    temp = tt
END FUNCTION
```

Chapter 9

Stellar Atmospheres

9.1 Introduction

In this chapter we will consider a simplified model of a stellar atmosphere. Such an atmosphere will be represented by a large table of data containing a number of physical quantities such as the temperature, the pressure, the density, the absorption coefficient, etc., at various levels in the atmosphere. The state of the matter in the atmosphere is different from the stellar interior state. The particles in the outer layers of a star are not completely ionized. There may still be a number of electrons surrounding the nucleus, a number which depends on the temperature and density. On the other hand, no nuclear energy production occurs since atmospheric temperatures and densities are far too low.

We must first clearly define what we mean by "atmosphere" of a star. This is quite simple and logical for most of the planets: the atmosphere is all the material outside the solid surface. A star, however, does not have a solid surface; thus, a more artificial definition of the concept is needed. There is no clear boundary between the atmosphere and the interior of the star.

The concept of effective temperature of a star is well known. It is the temperature used to plot the position of a star in the HR-diagram, and it is derived from a spectral analysis or from photometric observations at various wavelength bands. The effective temperature of our Sun is 5,800 Kelvin (K); for Sirius it is 10,000 K, and for Spica, 17,000 K. The temperature, density, and pressure all decrease from the stellar center to the surface. Hence, somewhere in the atmosphere the temperature is equal to the effective value. This level can be considered as an important reference layer of the atmosphere so that the atmosphere itself may be defined with respect to this layer. After we have introduced the concept of "optical depth," a more suitable and professional definition of the atmosphere will be given. The atmosphere thus defined will contain all the layers above the effective temperature reference layer and a smaller region beneath it.

Our models will be determined by two free parameters: the effective temperature T_{eff} and the gravitational acceleration at the "surface" of the star g_s. The physics described in this chapter can be applied for T_{eff} between 3,000 K and 30,000 K and for the log g_s between 2 and 4, so g_s is between 100 cm/s^2 and 10,000 cm/s^2. These two parameters are related to the mass M, the radius R, and the luminosity L of the star by the following formulae:

$$\log \frac{L}{L_\odot} = -15.05 + 4 \log T_{\text{eff}} + 2 \log \frac{R}{R_\odot} \tag{1}$$

$$g_s = 27538 \frac{M}{M_\odot} \left(\frac{R}{R_\odot} \right)^{-2}. \tag{2}$$

The relation between the mass and the luminosity of a star is expressed by a mass-luminosity relation. Different relations have been established for different types of stars. Giants, supergiants, and main-sequence stars all have their own mass-luminosity equations. These equations are based on both observations and theoretical stellar models, but are only first approximations. We have computed a number of models with a professional stellar model program and found for the main sequence the following relationship:

$$\log \frac{L}{L_\odot} = 3.34 \log \frac{M}{M_\odot}. \tag{3a}$$

This equation may be reproduced by any reader having successfully programmed the chapter on homogeneous stellar models. A factor 3.80 instead of 3.34 is then found as a result of the simplifications of our stellar models. The mass-luminosity relation for supergiants is as follows:

$$\log \frac{L}{L_\odot} = 1.17 + 2.62 \log \frac{M}{M_\odot}. \tag{3b}$$

Throughout the description of our model, we will assume a normal galactic chemical composition for the stellar material. For the outer layers containing the atmosphere, this means that 1 g of stellar material is composed of 0.70 g of hydrogen, 0.27 g of helium, and 0.03 g of other elements. Most of the formulae presented here are independent of this composition except for the absorption coefficient in which the galactic composition will be incorporated. Furthermore, the mean molecular weight is also a function of the chemical composition (and of the temperature and the density).

The model described here is certainly too simplified to be applicable in professional astronomy. We'll consider, for instance, the absorption coefficient κ as being a simple function of temperature and density. Especially in the atmosphere, κ should be computed very accurately, taking into account the various ionization stages of the chemical elements. Models to study the characteristics of stellar atmospheres are mostly very sophisticated. The problems involved become even more difficult if one wants to study and reproduce the detailed morphology of spectral lines. Other specialists are working on atmosphere models in which stellar wind is important, such as in stars with extended atmosphere and large mass loss rates (e.g., Wolf-Rayet stars). Hydrodynamic models are then to be used for the description of the stellar wind since the atmospheric particles move at considerable speeds. Generally less complicated models are used in programs for stellar evolution where the atmosphere is not the most important component of the problem.

9.2 Physical Background

Starting from the temperature, the density, and the pressure, we will now introduce and discuss the physical quantities needed in the equations of the atmosphere.

9.2.1 The Absorption Coefficient κ

Since the energy production of the star occurs in its interior, and the radiation—or in other words, the loss of energy—takes place at the stellar surface, there is a continuous flow of energy from the center to the surface. This energy flow consists of a large number of photons that may interact with the electrons and nuclei.

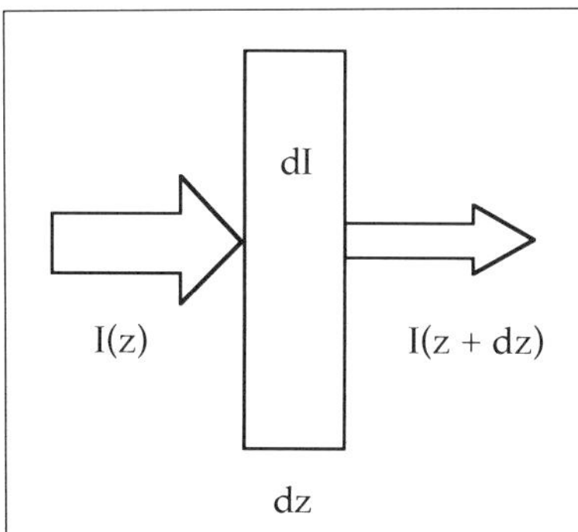

Fig. 9.1: *A thin layer of size dz absorbs a fraction dI of the incoming radiation flow $I(z)$. The resulting flow $I(z+dz)$ is smaller than $I(z)$.*

Consider now a thin layer at a certain height z in the atmosphere. It has a thickness dz and a density $\rho(z)$. Since the layer is very thin we may assume that the density and all other physical quantities are constant over the distance dz from the bottom to the top of the layer. Part of the radiation $I(z)$ entering the bottom of the layer will be absorbed by the material in the layer. Let us call this part $dI(z)$. For a sufficiently thin layer this fraction $dI(z)$ may be considered as being proportional to the amount of radiation entering the layer, to the thickness of the layer and to its density. Indeed, we may state as a first approximation that the layer twice as thick or twice as dense will absorb a double portion. Thus, $dI(z)$ is proportional to the product of $I(z)$, dz, and $\rho(z)$. The proportional factor is called the absorption coefficient κ:

$$dI(z) = -I(z)\,\kappa(z)\,\rho(z)\,dz. \tag{4}$$

Since temperature and density vary over the atmosphere, the absorption coefficient will be a function of the height. The formula contains a minus sign to express that there is a loss of radiation when dz is positive. In that way κ is a positive quantity.

Absorption is caused by a large variety of physical processes acting on various particles and elements in all possible degrees of ionization. The efficiency of these processes depends on the temperature and density of the material and of course on the chemical composition. Following are the three most important sources of absorption:

1. Bound-free transitions (photo-ionization). An atom absorbs a photon so that one of its electrons is able to escape.
2. Free-free transition. A free electron absorbs a photon to jump to a higher energy level while passing an atom.
3. Electron scattering. A process that doesn't depend on the wavelength.

The calculation of the absorption coefficient in an atmosphere is a very complicated problem for which various approximations have been computed. We will use one of the best known approximations—Kramer's law for opacity. For a galactic composition it is

$$\kappa = 1.984 \times 10^{24} \rho T^{-3.5} \mathrm{cm}^2/\mathrm{g}. \tag{5}$$

This formula enables us to calculate κ from any temperature and density. Electron scattering is not included in this approximation since it is only important at high temperatures.

9.2.2 The Optical Depth τ

The optical depth τ will be used as an independent variable in our model. All the other quantities (P, T, ρ,...) will be listed as functions of it. The optical depth τ is defined by

$$d\tau = -\kappa(z)\,\rho(z)\,dz. \tag{6}$$

The optical depth is, therefore, a dimensionless quantity. It increases when penetrating deeper and deeper in the atmosphere in the direction of the stellar center. An increase of the optical depth corresponds to a decrease of the vertical coordinate z. Let us now focus on its the physical interpretation. Combining (4) and (6) gives

$$dI = I d\tau. \tag{7}$$

When we integrate this formula between two values of τ we obtain

$$I_1 = I_0 e^{\tau_1 - \tau_0}. \tag{8}$$

This shows that over a layer with optical depth $d\tau = \tau_1 - \tau_0$, the radiation is decreased with a factor $\exp(-d\tau)$. This is the physical interpretation. A layer with a large absorption coefficient will absorb the radiation more efficiently than a layer with a small absorption coefficient. The former is called "optically thick"; the latter, "optically thin." A certain amount $d\tau$ in the first case corresponds to a smaller "real" distance dz in the atmosphere than for the thin case. This may be seen when we introduce the same $d\tau$ and the same density ρ but two different values of κ in (6). The resulting dz will also be different. Of course, the density plays also an important role in deciding whether the atmosphere is thin or thick. Thus, the larger the density and the absorption, the smaller dz for a fixed value of $d\tau$.

The reason that we use the optical depth instead of the height z as independent variable is now clear. The optical depth, or at least any change in optical depth, is really a characteristic of the stellar material since it depends on the density and the absorption coefficient. This is not the case, however, for the height z. A step of, for instance, 100 km in a supergiant with an extended atmosphere is something completely different from the same step in a main-sequence star. In the former, the physical quantities vary much less as a function of the height than in the latter. Large values of dz are needed to obtain significant changes of the physical quantities in a supergiant atmosphere. On the other hand, a certain change of the optical depth will yield comparable results for both atmospheres. Since a supergiant atmosphere is less dense, $d\tau$ will correspond there to a much larger distance step than in an MS-star.

The physical thickness dz of a layer with optical depth $d\tau$ is found from (6):

$$dz = -\frac{d\tau}{\kappa(z)\rho(z)}. \tag{9}$$

Since both $\kappa(z)$ and $\rho(z)$ may vary considerably in the atmosphere, dz also does if $d\tau$ is kept constant or slowly changes. We already explained that the optical depth increases inwards, although its zero level may be chosen at any point in the atmosphere. This increase occurs because only $d\tau$ has a physical meaning. We are now able to give a more accurate definition of the extent of the atmosphere:

- The level $\tau = 0$ is chosen at the outer boundary of the atmosphere.
- At that point the temperature, density, and pressure drop to zero.
- The bottom of the atmosphere is defined at the layer where $\tau = 1$. It is not possible to continue deeper because Kramer's law does not apply when τ is too high. Furthermore, convective layers may occur just at the base of the atmosphere. These cannot be handled with our simple method.
- The surface of the star is defined where the optical depth is 2/3. At that point the temperature will be equal to the effective temperature.

This definition is valid for every atmosphere with parameters within the ranges given in the introduction. A similar definition with dz would be impossible to give.

9.2.3 The Temperature T

In a simplified model, the temperature T is given as a function of the optical depth:

$$T^4 = \frac{3}{4} T_{\text{eff}}^4 \Big(\tau + q(\tau)\Big) \tag{10}$$

with

$$q(\tau) = 0.7104 - 0.1331 e^{-3.4488\tau}. \tag{11}$$

This formula, though not exact, is a good approximation for obtaining T. An exact formula as given by D. Mihalas in *Stellar Atmospheres* can only be written with integrals. For the top, surface, and bottom we find the following:

$$\text{at } \tau = 0: \quad T = 0.81 T_{\text{eff}}$$

$$\text{at } \tau = \frac{2}{3}: \quad T = 1.00 T_{\text{eff}}$$

$$\text{at } \tau = 1: \quad T = 1.06 T_{\text{eff}}$$

The temperature will vary through the atmosphere over a range of 0.81 to 1.06 times the effective temperature. For the Sun, for instance, this means from 4,700 K to 6,150 K for an effective temperature of 5,800 K.

9.2.4 The Gas Pressure P_g and the Mean Molecular Weight μ

These two quantities are related to the temperature and the density by the equation of state for an ideal gas

$$P_g = \frac{\rho R T}{\mu} \tag{12}$$

where R is the gas constant

$$R = 8.314 \times 10^7 \text{ erg/K mole.}$$

Similar to the absorption coefficient, the mean molecular weight is not a simple, straightforward function of the temperature and density. It varies with these two parameters as a consequence of the effect of partial ionization, which depends critically on T and ρ. In the stellar interior, where all matter is completely ionized, it has a simple expression, but in the atmosphere things are so complicated that we may only proceed by making an additional and dangerous assumption: we will take μ constant in the whole atmosphere. Even the argument that T varies over a range of only 0.81 to 1.06 does not justify this assumption since there is still the variation of the density. Nevertheless, we will use mean values of μ computed with professional models and list them as a function of the initial parameters T_{eff} and $\log g_s$ (see below). The exact definition of the mean molecular weight is

$$\mu^{-1} = \sum \frac{n_i X_i}{A_i}$$

in which X_i is the fraction by weight of the chemical element i; A_i, its atomic weight (Carbon = 12); and n_i, the number of free particles per atom of element i. For instance, when helium is totally ionized, $n = 3$—namely, the naked nucleus and the two free electrons. The summation is performed over all the elements present in the stellar material, taking into account their presence by means of their fraction by weight X in 1 g of material.

Let us consider a mixture that only contains hydrogen and helium. Suppose also that 1 g of material contains 0.70 g hydrogen (so $X_1 = .70$) and 0.30 g helium ($X_2 = .30$). Furthermore, we have $A_1 = 1$ (the lonely proton in the hydrogen nucleus) and $A_2 = 4$ (two protons and two neutrons in a helium nucleus).

- When none of the elements is ionized, $(n_1 = n_2 = 1) : \mu = 1.29$.
- When only hydrogen is ionized, $(n_1 = 2) : \mu = 0.678$.
- When hydrogen is ionized, but helium only partially (i.e., one free and one bound electron), $n_1 = 2; n_2 = 2 : \mu = 0.645$.

- In the case of complete ionization, we have $n_1 = 2$ (one H-nucleus plus its free electron) and $n_2 = 3$ (see above). This gives $\mu = 0.615$.

The mixture considered in our atmosphere model differs from the mixture used in this example. The fraction by weight of the heavier elements is 0.03.

Table 9-1
Average mean molecular weight used in our models.

$T_{\rm eff}$ =	5,000	6,000	7,000	8,000	10,000	12,000
$\log g_s$						
4	1.289	1.287	1.285	1.277	1.048	0.724
3	1.287	1.285	1.284	1.266	0.868	0.688
2	1.285	1.285	1.280	1.232	0.738	0.675

$T_{\rm eff}$ =	14,000	16,000	18,000	20,000	25,000	30,000
$\log g_s$						
4	0.682	0.672	0.666	0.656	0.641	0.640
3	0.673	0.659	0.659	0.646		

The largest jump in the table occurs near a temperature of 10,000 K, the temperature above which hydrogen is ionized. This means that for every hydrogen atom the number of free particles is doubled from one to two. The new free particles are the electrons that were bound to the nucleus at lower temperatures. When they become free, the material contains a larger fraction of less massive particles (the mass of an electron is only $1/1830$ of the mass of a proton or neutron), thus reducing the mean weight of a mole of particles. This ionization does not occur suddenly, but takes place between 8,000 K and 12,000 K. The ionization of helium is less spectacular since the abundance of helium is only 0.27, and much higher temperatures are needed. Furthermore, helium has two electrons so that the transition from neutral helium to complete ionization occurs in two steps.

9.2.5 The Radiation Pressure P_r

The expression for radiative pressure in an atmosphere in radiative equilibrium is

$$P_r = \frac{1}{3} a T^4 \tag{13}$$

where

$$a = 7.56464\ 10^{-15}\,\mathrm{erg/cm^3}/K^4.$$

The radiation pressure is generated by the radiation field. Particles may absorb and emit energy. For emission there is no preferential direction. Such a process (with no preferential direction) is called an *isotropic process*. Absorption, however, requires a radiation field. In our case, the energy flow is directed outwards; consequently, there are more photons to be absorbed at the lower side of a particle than at the upper side. Every absorption also means a little gain in momentum. When the photon "hits" the particle, it also gives the particle a little push. Since there are more absorptions at the base, the particles will be pushed outwards by the radiation field.

The radiation pressure depends only on the temperature and will therefore decrease outwards. This process is only important for sufficiently high values of T.

9.2.6 The Equation of Hydrostatic Equilibrium

This equation describes the balance between the gravitational field and the outward decrease of pressure in the atmosphere. When these two forces cancel each other, the stellar material is at rest, at *hydrostatic equilibrium*. A complete deduction of this important formula is presented in the chapter on polytropes. Its general form is

$$\frac{dP}{dr} = -G\rho(r)\frac{M_r}{r^2} \tag{14}$$

where

P is the total pressure at distance r from the stellar center,

$\rho(r)$ is the density at that point,

M_r is the amount of mass inside a sphere with radius r,

G is the gravitational constant, and

r is the radial coordinate, generally called z in the atmosphere.

For the atmosphere, M_r and r may be replaced by the total mass and radius because the atmospheres considered here really contain a negligible fraction of the mass, while the total size is small compared with the radius of the star. With the relation (2) between the surface gravitational acceleration, the total mass, and the radius, and the definition of the optical depth and the absorption coefficient (9), we may write

$$dP = \frac{g_s}{\kappa}d\tau. \tag{15}$$

In this equation the pressure P is the total pressure, i.e., the combined effect of gas and radiation pressure.
Thus

$$dP_g = \frac{g_s}{\kappa}d\tau - dP_r. \tag{16}$$

The outward acceleration g_r due to radiation pressure is now defined in the same way as the gravitational acceleration:

$$dP_r = \frac{g_r}{\kappa}d\tau \tag{17}$$

which means

$$g_r = \kappa \frac{dP_r}{d\tau}.$$

If we now replace P_r by its definition (13) and then T^4 by (10,11), we have

$$g_r = \frac{\kappa a T_{\text{eff}}^4}{4} \left(1 + 0.459 e^{-3.4488\tau}\right). \tag{18}$$

The effective surface acceleration is finally defined by

$$g_e = g_s - g_r. \tag{19}$$

This acceleration is a function of the absorption coefficient and therefore of the temperature and density. It is, in fact, the gravitational acceleration with a correction for the effects of radiation pressure. It is inserted in the equation of hydrostatic equilibrium by means of (16) so that this equation finally becomes

$$dP_g = \frac{g_e}{\kappa} d\tau. \tag{20}$$

This form of the equation of hydrostatic equilibrium will be used in our model. Since it is a differential equation, we'll have to solve it numerically as a function of the optical depth. We will start at the outer boundary of the atmosphere—where the optical depth is zero and most of the physical quantities such as T and P are known—and proceed inwards until the optical depth becomes one.

9.2.7 Summary of the Equations

The physical meaning of all the relevant formulae of our atmospheric model has now been discussed. The model is hence the result of solving the problems that follow.

Independent variable:

$$\text{optical depth } \tau \tag{21}$$

Temperature T:

$$T^4 = \frac{3}{4} T_{\text{eff}}^4 \left(\tau + q(\tau)\right) \tag{22}$$

$$q(\tau) = 0.7104 - 0.1331 e^{-3.4488\tau} \tag{23}$$

Radiation pressure:

$$P_r = \frac{1}{3} a T^4 \tag{24}$$

Effective acceleration:

$$g_e = g_s - g_r \tag{25}$$

$$g_r = \frac{\kappa a T_{\text{eff}}^4}{4}\left(1 + 0.459e^{-3.4488\tau}\right) \tag{26}$$

Absorption coefficient:

$$\kappa = 1.984 \times 10^{24} \rho T^{-3.5} \tag{27}$$

Equation of state:

$$P_g = \frac{\rho R T}{\mu} \tag{28}$$

Hydrostatic equilibrium:

$$dP_g = \frac{g_e}{\kappa} d\tau \tag{29}$$

9.3 Numerical Method

First, the two free parameters T_{eff} and g_s are to be selected. Then the corresponding value of μ is selected from the table. Finally, we may start the iterations.

9.3.1 The Initial Values at $\tau = 0$

At this starting point, the temperature is found with (22,23) by taking $\tau = 0$ and then the radiation pressure from (24). Four other initial values are to be selected, namely for P_g, κ, ρ, and g_e. Since we have only three equations at our disposal, one of the four may be chosen freely. Notice that equation (29) is a differential equation and therefore not suited to compute initial values. The most logical choice is to take the density at the boundary zero. However, it is not possible to do so since then κ would also be zero, thus making the denominator of (29) zero. This problem may be solved by starting from a very small density at the surface. Test calculations have shown that 10^{-13} is a reasonable choice. In this way, the solutions are sufficiently independent of the initial value of the density, and no difficulties in (29) are encountered.

With this choice for the density, the initial values of P_g, κ, and g_e are found from (28,27,26,25). In this way, we obtain the first line of the table representing the stellar atmosphere:

$$\tau = 0: \;\; T_0, \;\; P_{g,0}, \;\; P_{r,0}, \;\; \kappa_0, \;\; \rho_0, \;\; g_{e,0}.$$

9.3.2 Proceeding from a Level τ_i to τ_{i+1}

Special attention has to be paid at the choice of the step-size $d\tau$, as discussed in subsection 3.3. Let us take $d\tau_i$ as notation for the step-size from τ_i to τ_{i+1}. Thus

$$d\tau_i = \tau_{i+1} - \tau_i.$$

Suppose that the state at level i is known. Let us write the values of the physical quantities at that level with index i. We then proceed to the new level $i+1$ by means of the midpoint method in the following way.

First:

$$\tau_{i+1/2} = \tau_i + \frac{1}{2} d\tau_i \tag{30}$$

$$T_{i+1/2} = \text{from } (22, 23), \text{ with } \tau_{i+1/2} \tag{31}$$

$$P_{g,i+1/2} = P_{g,i} + 0.5 d\tau_i \frac{g_{e,i}}{\kappa_i} \tag{32}$$

$$\rho_{i+1/2} = \text{from } (28), \text{ every variable evaluated at } i+1/2 \tag{33}$$

$$\kappa_{i+1/2} = \text{from } (27), \text{ every variable evaluated at } i+1/2 \tag{34}$$

$$g_{e,i+1/2} = \text{from } (25,\ 26), \text{ every variable evaluated at } i+1/2 \tag{35}$$

Then:

$$\tau_{i+1} = \tau_i + d\tau_i \tag{36}$$

$$T_{i+1} = \text{from } (22,\ 23), \text{ with } \tau_{i+1} \tag{37}$$

$$P_{g,i+1} = P_{g,i} + d\tau_i \frac{g_{e,i+1/2}}{\kappa_{i+1/2}} \tag{38}$$

$$\rho_{i+1} = \text{from } (28), \text{ every variable evaluated at } i+1 \tag{39}$$

$$\kappa_{i+1} = \text{from } (27), \text{ every variable evaluated at } i+1 \tag{40}$$

$$g_{e,i+1} = \text{from } (25,\ 26), \text{ every variable evaluated at } i+1 \tag{41}$$

The real distance between the two levels is calculated from

$$dz = \frac{d\tau_i}{(\kappa_{i+1/2})\,(\rho_{i+1/2})}. \tag{42}$$

The minus sign is omitted since dz is now used as a distance instead of a vertical coordinate. The distance step dz will vary over an atmosphere and from one atmosphere to another, between a few hundred and a few thousand kilometers. The density and absorption coefficients are evaluated in the middle of the step, corresponding with the philosophy of the midpoint method. The zero level for z may be chosen anywhere. We'll place it at the outer edge of the atmosphere where the optical depth is zero. The total size of the atmosphere is found by adding the dz values of all the layers.

9.3.3 The Step-size $d\tau_i$

The calculation of an atmospheric model is performed with a variable step-size because the physical quantities vary too much to allow constant steps. It is better for the first steps near the outer edge of the atmosphere to be small. As penetration deepens, the step-size may gradually be increased. An elegant and easy way to select a suitable step-size is to take $d\tau_i$ proportional to τ_i. Only the first step from τ_0 to τ_1 has a fixed value since τ_0 equals zero.

Take

$$d\tau_0 = 0.001.$$

From then on

$$d\tau_i = 0.25\tau_i$$

which implies that

$$\tau_{i+1} = 1.25\tau_i.$$

With these relations it is possible to compute the values of τ_i which will be used during the iteration procedure:

$$\begin{aligned} \tau_0 &= 0.00000 \text{ and } d\tau_0 = 0.001 \text{ (fixed value)} \\ \tau_1 &= 0.00100 \text{ and } d\tau_1 = 0.00025 \\ \tau_2 &= 0.00125 \text{ and } d\tau_2 = 0.0003125, \text{ etc.} \end{aligned}$$

In this way we integrate deeper and deeper until the bottom level is reached ($\tau > 1$). How many layers are needed to arrive there?

Since

$$\tau_i = \tau_{i-1} + d\tau_{i-1} = \tau_{i-1} + 0.25\tau_{i-1} = 1.25\tau_{i-1}$$

we may apply this $i-1$ times:

$$\tau_i = (1.25)^{i-1}\tau_1 = 0.001(1.25)^{i-1}.$$

Suppose that N steps are needed to arrive at a level where $\tau > 1$. Then

$$1 = 0.001(1.25)^{N-1} \text{ or } 1000 = (1.25)^{N-1}.$$

Taking logarithms gives

$$3 = (N-1)\log 1.25 \text{ or } N = 32.$$

The optical depth at the bottom of the atmosphere is $0.001 \times (1.25)^{31}$, which gives 1.00974. The complete list of optical depths encountered during the iteration is:

(0)	0.00000	(11)	0.00931	(22)	0.10842
(1)	0.00100	(12)	0.01164	(23)	0.13553
(2)	0.00125	(13)	0.01455	(24)	0.16941
(3)	0.00156	(14)	0.01819	(25)	0.21176
(4)	0.00195	(15)	0.02274	(26)	0.26470
(5)	0.00244	(16)	0.02842	(27)	0.33087
(6)	0.00305	(17)	0.03553	(28)	0.41359
(7)	0.00381	(18)	0.04441	(29)	0.51699
(8)	0.00477	(19)	0.05551	(30)	0.64623
(9)	0.00596	(20)	0.06939	(31)	0.80779
(10)	0.00745	(21)	0.08674	(32)	1.00974

In each of the 33 levels we compute the following quantities:

$$\tau, \; T, \; P_g, \; P_r, \; \rho, \; \kappa, \; g_e, \; z.$$

The total output of our program is a table with $8 \times 33 = 264$ data, representing a stellar atmosphere.

9.4 Some Remarks Concerning the Program

Using subprograms as much as possible (`Gosub` in Basic, `Subroutine` in Fortran, `Procedure` in Pascal, ...) will simplify writing the program since you may use them later in your main program. Furthermore, the use of subprograms will shorten your program since most of the formulae are needed two times during each step of the integration. Using subprograms is especially important for programmable pocket calculators with their rather limited memory size. PC-owners have more liberty, but using subprograms will still make your program easier to understand when you read it later. (See flowchart on following page.) We suggest the following subprograms—five functions and two procedures (or subroutines).

Function `TEMP` computes the temperature for a given optical depth (22,23).

Function `ABSORP` computes the absorption coefficient for a given density and temperature from (27).

Function `RADPRESS` computes the radiation pressure for a given temperature (24).

Function `GEFFECT` computes the effective gravitational acceleration from g_s and κ (25,26).

Function `DENS` computes the density for given T and P from (28).

Procedure `DISP` displays the values of τ, T, P_g, P_r, ρ, κ, g_e, dz.

Procedure `STAP` computes the values at layer $i+1$ from layer i:

$$
\begin{aligned}
\tau_{i+1/2} &= \tau_i + \tfrac{1}{2}\, d\tau_i \\
T_{i+1/2} &= \texttt{TEMP}\left(\tau_{i+1/2}\right) \\
P_{g,i+1/2} &= P_{g,i} + 0.5 d\tau_i \frac{g_{e,i}}{\kappa_i} \\
\rho_{i+1/2} &= \texttt{DENS}\left(T_{i+1/2}, P_{g,i+1/2}\right) \\
\kappa_{i+1/2} &= \texttt{ABSORP}\left(\rho_{i+1/2}, T_{i+1/2}\right) \\
g_{e,i+1/2} &= \texttt{GEFFECT}\left(g_s, \kappa_{i+1/2}, \tau_{i+1/2}\right) \\
\tau_{i+1} &= \tau_i + d\tau_i \\
T_{i+1} &= \texttt{TEMP}\left(\tau_{i+1}\right) \\
P_{g,i+1} &= P_{g,i} + d\tau_i \frac{g_{e,i+1/2}}{\kappa_{i+1/2}} \\
\rho_{i+1} &= \texttt{DENS}\left(T_{i+1}, P_{g,i+1}\right) \\
\kappa_{i+1} &= \texttt{ABSORP}\left(\rho_{i+1}, T_{i+1}\right) \\
g_{e,i+1} &= \texttt{GEFFECT}\left(g_s, \kappa_{i+1}, \tau_{i+1}\right) \\
dz &= (42);\; P_r = \texttt{RADPRESS}\left(T_{i+1}\right)
\end{aligned}
$$

The main program becomes rather simple with these subprograms. It contains three parts. First, the input values and initial conditions are computed. Then the first step is executed separately from the other steps, because of the difference in the method used to select the step-size $d\tau$. The third part is the main cycle, which is executed until the optical depth exceeds 1.

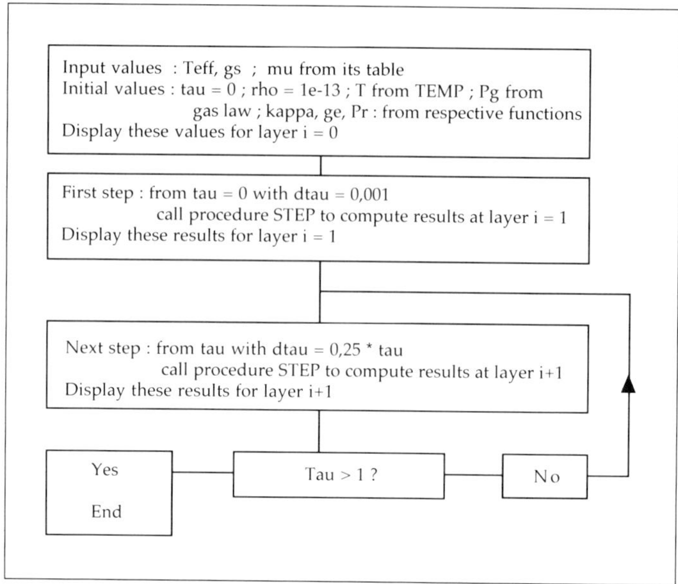

9.5 Applications

As a first example, we give in Table 9-1 the complete result for a model with an effective temperature of 10,000 K and a surface gravitational acceleration of $\log g_s = 4$. This means that $\mu = 1.048$. The initial density is 10^{-13}.

The density is expressed in units of 10^{-11}g/cm^3. The temperature is in Kelvin. All other quantities are expressed in centimeter-gram-second units. To save space, only a limited number of digits are presented. Truncation errors may slightly change the results when going deeper in the atmosphere. The total size of the atmosphere is about 3,400 km.

The atmosphere of this star corresponds to a main-sequence star in the HR-diagram. Using the general formulae (1,2,3a), we are able to derive some properties of this star as:

$$\text{Mass} = 3.9M_\odot$$

$$\text{Radius} = 3.2R_\odot$$

$$\text{Luminosity} = 93L_\odot$$

As mentioned before, this model lies on the main sequence. The temperature corresponds to a spectral type A0. A typical example of such a star is Sirius.

Table 9-1

τ	T	P_g	P_r	ρ	κ	g_e	z
0.00000	8111.8	0.06	10.92	0.010	0.00413	9999.89	0.0
0.00100	8116.9	0.17	10.95	0.026	0.01059	9999.71	0.1
0.00125	8118.2	0.49	10.95	0.076	0.03130	9999.14	1.9
0.00156	8119.8	1.45	10.96	0.225	0.09268	9997.45	14.3
0.00195	8121.8	4.16	10.97	0.646	0.26526	9992.70	92.0
0.00244	8124.2	9.88	10.98	1.533	0.62937	9982.68	368.8
0.00305	8127.3	16.38	11.00	2.540	1.04144	9971.36	654.1
0.00381	8131.2	22.36	11.02	3.466	1.41833	9961.03	847.1
0.00477	8136.0	28.18	11.05	4.367	1.78341	9951.04	994.1
0.00596	8142.1	34.14	11.08	5.286	2.15312	9940.97	1116.8
0.00745	8149.5	40.40	11.12	6.249	2.53736	9930.55	1225.2
0.00931	8158.9	47.10	11.17	7.277	2.94297	9919.60	1324.3
0.01164	8170.4	54.36	11.24	8.387	3.37518	9908.02	1417.3
0.01455	8184.8	62.31	11.32	9.596	3.83815	9895.73	1506.1
0.01819	8202.6	71.07	11.41	10.922	4.33518	9882.67	1592.1
0.02274	8224.5	80.78	11.54	12.381	4.86870	9868.85	1676.1
0.02842	8251.6	91.61	11.69	13.992	5.44016	9854.30	1759.0
0.03553	8284.9	103.74	11.88	15.783	6.04973	9839.13	1841.4
0.04441	8325.6	117.38	12.11	17.772	6.69612	9823.49	1923.8
0.05551	8375.1	132.80	12.41	19.988	7.37630	9807.63	2006.8
0.06939	8435.2	150.32	12.77	22.464	8.08531	9791.85	2090.7
0.08674	8507.4	170.33	13.21	25.237	8.81624	9776.53	2176.2
0.10842	8593.8	193.29	13.75	28.352	9.56039	9762.10	2263.6
0.13553	8696.2	219.81	14.42	31.862	10.3077	9749.00	2353.6
0.16941	8816.5	250.61	15.24	35.831	11.0476	9737.61	2446.8
0.21176	8956.6	286.62	16.23	40.338	11.7695	9728.20	2543.6
0.26470	9118.2	328.97	17.43	45.477	12.4638	9720.87	2644.9
0.33087	9303.1	379.08	18.89	51.363	13.1217	9715.46	2751.1
0.41359	9513.1	438.74	20.65	58.135	13.7356	9711.60	2863.1
0.51699	9750.2	510.18	22.79	65.957	14.2975	9708.74	2981.4
0.64623	10016.8	596.21	25.39	75.028	14.7985	9706.31	3106.8
0.80779	10315.9	700.41	28.56	85.585	15.2286	9703.85	3240.3
1.0097	10651.2	827.37	32.45	97.916	15.5776	9701.25	3382.6

A second example is a model with the same effective temperature of 10,000 K but with $\log g_s = 2$. The corresponding value of $\mu = 0.738$. Results are shown in Table 9-2.

The total size of the atmosphere is now more than 500,000 km. This differs greatly from the first example, although both have the same effective temperature. Clearly this atmosphere belongs to a supergiant with extended layers compared to a main-sequence star.

A typical star of this kind is Deneb, with an effective temperature of 9,600 K and $\log g_s = 2.2$. From our general formulae (1,2,3b), we find the following:

$$\text{Mass} = 23M_\odot$$

$$\text{Radius} = 79R_\odot$$

$$\text{Luminosity} = 55000L_\odot$$

Table 9-2

τ	T	P_g	P_r	ρ	κ	g_e	z
0.00000	8111.8	0.09	10.92	0.010	0.00413	99.89	0.0
0.00100	8116.9	0.25	10.95	0.027	0.01106	99.70	1363.5
0.00125	8118.2	0.64	10.95	0.070	0.02896	99.20	28339.4
0.00156	8119.8	1.22	10.96	0.134	0.05505	98.48	74070.9
0.00195	8121.8	1.77	10.97	0.193	0.07931	97.82	106182
0.00244	8124.2	2.28	10.98	0.249	0.10220	97.19	129484
0.00305	8127.3	2.79	11.00	0.305	0.12502	96.56	148376
0.00381	8131.2	3.32	11.02	0.363	0.14850	95.92	164771
0.00477	8136.0	3.89	11.05	0.424	0.17316	95.25	179620
0.00596	8142.1	4.49	11.08	0.489	0.19938	94.53	193456
0.00745	8149.5	5.14	11.12	0.560	0.22745	93.77	206613
0.00931	8158.9	5.86	11.17	0.637	0.25766	92.96	219311
0.01164	8170.4	6.64	11.24	0.721	0.29023	92.09	231708
0.01455	8184.8	7.50	11.32	0.813	0.32534	91.16	243922
0.01819	8202.6	8.45	11.41	0.915	0.36314	90.17	256047
0.02274	8224.5	9.51	11.54	1.026	0.40370	89.13	268164
0.02842	8251.3	10.69	11.69	1.150	0.44704	88.03	280345
0.03553	8284.9	12.01	11.88	1.286	0.49306	86.89	292657
0.04441	8325.6	13.48	12.11	1.437	0.54156	85.72	305168
0.05551	8375.1	15.14	12.41	1.605	0.59220	84.56	317947
0.06939	8435.2	17.02	12.77	1.791	0.64452	83.41	331070
0.08674	8507.4	19.15	13.21	1.998	0.69793	82.31	344616
0.10842	8593.8	21.58	13.75	2.229	0.75176	81.29	358671
0.13553	8696.2	24.39	14.42	2.489	0.80528	80.39	373330
0.16941	8816.5	27.63	15.24	2.782	0.85776	79.63	388693
0.21176	8956.6	31.42	16.23	3.114	0.90857	79.02	404863
0.26470	9118.2	35.87	17.43	3.492	0.95715	78.56	421947
0.33087	9303.1	41.15	18.89	3.926	1.00305	78.25	440055
0.41359	9513.1	47.44	20.65	4.427	1.04587	78.04	459294
0.51699	9750.2	54.98	22.79	5.006	1.08511	77.90	479778
0.64623	10016.8	64.28	25.39	5.679	1.11013	77.77	501626
0.80779	10315.9	75.12	28.56	6.464	1.14012	77.64	524975
1.00974	10651.2	88.57	32.45	7.381	1.17426	77.48	549985

The same atmosphere computed with an initial density of 10^{-12} instead of 10^{-13} results at the bottom in $P_g = 88.599$ and $\rho = 7.381$—which differs only marginally from the data in the table above. This small difference shows the relative freedom in selecting the initial density.

9.6 The Program Listing

The following sample program is written in MicroSoft QuickBasic 4.5:

```
DECLARE FUNCTION radpress (t)
DECLARE FUNCTION temp (tau, teff)
DECLARE FUNCTION absorp (t, d)
DECLARE FUNCTION dens (t, p, mu)
DECLARE FUNCTION geffect (gs, teff, k, tau)
DECLARE SUB stap (teff, gs, mu, dtau, tau, t, pg, d, kk, ge, kk1, d1)
DECLARE SUB disp (i%, tau, t, pg, pr, rho, kappa, ge, z)
```

```
COMMON SHARED pi, g, a, r

'some physical constants used in the program :
pi = 3.1415926535#
g = 6.673E-08
a = 7.56464E-15
r = 8.314E+07

CLS
PRINT " "
PRINT "Astrophysics With a PC : STELLAR ATMOSPHERE"
PRINT "-------------------------------------------------"
PRINT " "
PRINT " ----------- Minimal solution program -----------"
PRINT " "
PRINT "Input of initial conditions and model parameters : "
INPUT "Effective temperature         : ", teff
INPUT "Surface gravit.accel.(log10)  : ", gs
INPUT "Average mean molecular weight : ", mu
gs = 10 ^ gs


' display heading of main table
PRINT " "
PRINT "  i    tau      T      Pg      Pr      d       k        ge        z(km)"


' compute central value of physical quantities
tau = 0
d = 1E-13
t = temp(tau, teff)
pg = r * d * t / mu
pr = radpress(t)
kk = absorp(t, d)
ge = geffect(gs, teff, kk, tau)
z = 0
i% = 0
CALL disp(i%, tau, t, pg, pr, d, kk, ge, z)


'compute physical quantities in the layer 1
i% = 1
dtau = .001
CALL stap(teff, gs, mu, dtau, tau, t, pg, d, kk, ge, kk1, d1)
z = dtau / kk1 / d1
CALL disp(i%, tau, t, pg, pr, d, kk, ge, z)


' next FOR-cycle computes the other layers
FOR i% = 2 TO 32
    dtau = .25 * tau
    CALL stap(teff, gs, mu, dtau, tau, t, pg, d, kk, ge, kk1, d1)
    dz = dtau / kk1 / d1
    pr = radpress(t)
    z = z + dz
    CALL disp(i%, tau, t, pg, pr, d, kk, ge, z)


'  pause after having shown layer 15
    IF i% = 15 THEN
        PRINT "press any key to continue"
        DO
        LOOP WHILE INKEY$ = ""
    END IF
NEXT i%
```

```
PRINT " "
PRINT "Model complete. Press any key to stop the program"
DO
LOOP WHILE INKEY$ = ""
END


FUNCTION absorp (t, d)
'
' computes the absorption coefficient for given temperature T and density d
'
    absorp = 1.984E+24 * d / t ^ 3.5
END FUNCTION


FUNCTION dens (t, p, mu)
'
' computes the density for input values of temperature T, pressure P and mu
'
    dens = p * mu / t / r
END FUNCTION


SUB disp (i%, tau, t, pg, pr, rho, kk, ge, z)
'
' this SUB is called each layer to display the results for that layer
'
    PRINT USING "### ##.##### #####.# ###.## ###.## ###.### ##.#####_
 #####.## #######.#"; i%; tau; t; pg; pr; rho * 1E+11; kk; ge; z / 100000!
END SUB


FUNCTION geffect (gs, teff, k, tau)
'
'  computes the effective gravitational acceleration
'
    geffect = gs - k * a * teff ^ 4 / 4 * (1 + .459 * EXP(-3.4488 * tau))
END FUNCTION


FUNCTION radpress (t)
'
' computes the radiation pressure for given temperature T and density d
'
     radpress = a / 3 * t ^ 4
END FUNCTION


SUB stap (teff, gs, mu, dtau, tau, t, pg, d, kk, ge, kk1, d1)
'
' computes the physical quantities at layer i+1 starting from layer i
'
    tau1 = tau + .5 * dtau
    t1 = temp(tau1, teff)
    pg1 = pg + .5 * dtau * ge / kk
    d1 = dens(t1, pg1, mu)
    kk1 = absorp(t1, d1)
    ge1 = geffect(gs, teff, kk1, tau1)
    tau = tau + dtau
    t = temp(tau, teff)
    pg = pg + dtau * ge1 / kk1
    d = dens(t, pg, mu)
```

```
    kk = absorp(t, d)
    ge = geffect(gs, teff, kk, tau)
 END SUB


FUNCTION temp (tau, teff)
'
' computes the temperature for given optical depth tau in effective_
 temperature teff
'
     q = .7104 - .1331 * EXP(-3.4488 * tau)
     temp = teff * (.75 * (tau + q)) ^ .25
END FUNCTION
```

Chapter 10

The Structure of White Dwarfs

10.1 Introduction

Up until now a few thousand white dwarfs have been discovered. They are located in the Hertzsprung-Russell diagram in a region well beneath the main sequence in the spectral classes A and F. One of the best known white dwarfs is the companion of the brightest star for terrestrial observers: Sirius. Its faint companion Sirius B, which was discovered in 1862, is sometimes visible in sufficiently large amateur telescopes. The problem is not the weak magnitude of the B-component, but the large magnitude contrast with the bright Sirius A. Sirius B is a white dwarf with a mass of about $0.9M_\odot$ and an absolute magnitude of +11.5. Both components of the Sirius binary have an effective temperature of about 10,000 K.

White dwarfs are the final stage in the evolution of low mass stars with initial masses smaller than $3M_\odot$. Our Sun will eventually evolve into such a state, in which all energy production has stopped. The small amount of energy radiated away in space and responsible for the faint magnitude is supplied by the thermal energy still resident in the star from earlier evolutionary stages. Since this energy supply is limited, the white dwarf will slowly become fainter and will finally fade out. This process takes billions of years but leads to an inevitable end. Estimates conclude that there should be about 10 billion inactive white dwarfs in our galaxy.

The position of white dwarfs in the HR-diagram shows that they have a very small radius compared to normal main-sequence stars. On the other hand, the masses of white dwarfs are in the order of one solar mass. These two facts show that these stars have very large densities. The average density of a white dwarf is about 10^6 g/cm^3, compared to only 1.42 for the Sun. The central densities of white dwarfs are even higher. The physical state of the matter in the interior of white dwarfs can only be explained in terms of electron degeneracy. This state allows extremely high densities and pressures. Some "normal" terrestrial laws of physics, such as the perfect gas law, no longer hold and should be replaced by an appropriate law for degenerate matter. An excellent and detailed description of the mathematical deduction of the formulae for degenerate matter may be found in *Structure and Evolution of the Stars* by Schwarzschild or in *Principles of Nucleosynthesis* by Clayton. Important theoretical work on white dwarfs was performed by Chandrasekhar.

10.2 Electron Degeneracy

The particles of a gas, for instance the free electrons in the stellar interior, do not all have the same velocity. The majority of the velocities is clustered around some average value, but some particles may have a lower or higher velocity. Under normal terrestrial or stellar temperatures and densities, the velocity distribution, i.e., the number of particles having a certain speed v, is a Maxwell-Boltzmann distribution, a clock-shaped function, shifting towards higher velocities as the temperature increases. Since the momentum of a particle is defined as the product of its mass and its velocity, the momentum distribution has the same shape.

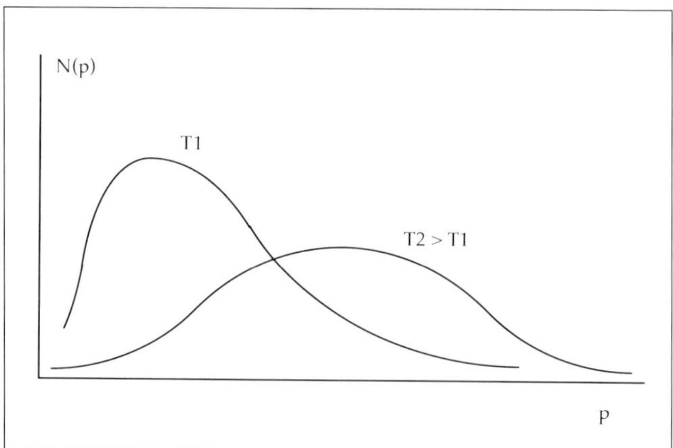

Fig. 10.1: *Maxwell-Boltzmann distributions.*

If we increase the density, keeping the temperature constant, the distribution is increased proportionally, but stays in the same momentum region. However, as the density becomes extremely high, an important physical process creates deviations from the normal Maxwell-Boltzmann distribution. This process is a consequence of the exclusion principle of Pauli: the maximum number of electrons in a volume dV and with momentum between p

and $p + dp$ is limited by

$$N_{e,\max}\, dV\, dp_x dp_y dp_z = \frac{2}{h^3} dV\, dp_x dp_y dp_z. \tag{1}$$

Although this upper limit corresponds to a considerable number of electrons, the densities needed to reach this limit are present in some types of stars. Under normal circumstances the top of the Maxwell-Boltzmann distribution is far below the limit value, but in the interior of white dwarfs, this limit is attained. We therefore say that the interior of white dwarfs is *electron degenerate*. For the even higher densities of neutron stars, the distribution of atomic nuclei is affected in a similar way, which is then called *ion degeneracy*.

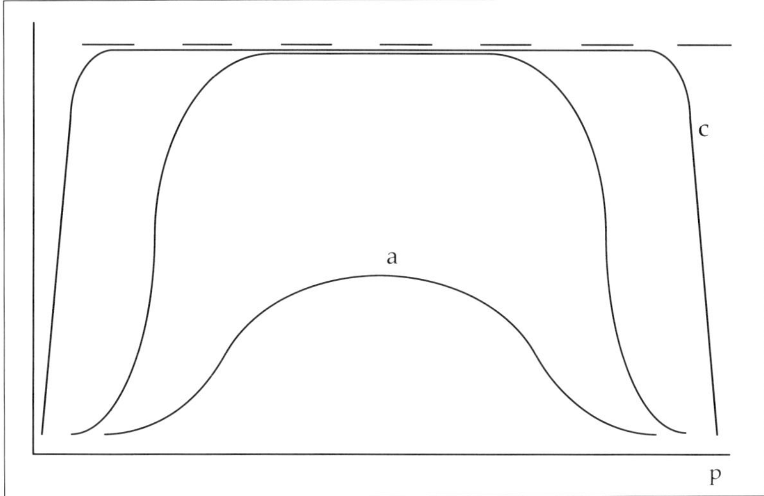

Fig. 10.2: *From a Maxwell-Boltzmann distribution to electron degeneracy.*

What does this electron degeneracy mean for the momentum distribution of the electrons? In Fig. 10.2 we have plotted three distributions for densities a, b, and c. For density a, the lowest one, the distribution is still Maxwellian since its top is far below the limit. At density b, however, the top would exceed the maximum allowed number of electrons. Therefore, the electrons above this maximum are now located at the slopes of the ideal distribution. These slopes become steeper. If we keep increasing the density to level c, we finally end up with an almost rectangular distribution in which all the volumes and all the momentum values between zero and a certain maximum momentum p are filled with the maximum number. We then call the matter *completely electron degenerate*.

Degenerate matter obeys its own physical laws. These laws have been studied profoundly by various astronomers so that it is now possible to use a generally accepted and known basic theory. In particular, the equation of state is no longer the well known ideal gas law. For electron degeneracy, the new law may be written using a dimensionless parameter:

$$x = \frac{p_o}{m_e c} \tag{2}$$

where

m_e = mass of the electron;
c = speed of light;
p_o = maximum momentum of the degenerate momentum distribution.

Deriving the equation of state is mathematically rather complicated and may be found in most books on stellar structure. We only present the result of the computations. The pressure P is given by

$$P = 6.01 \times 10^{22} f(x) \tag{3}$$

with

$$f(x) = x\left(2x^2 - 3\right)\left(1 + x^2\right)^{1/2} + 3\ln\left(x + \left(1 + x^2\right)^{1/2}\right). \tag{4}$$

where "ln" stands for the natural logarithm. The density is then given by

$$\rho = 9.82 \times 10^5 \mu_e\, x^3 \tag{5}$$

with the mean molecular weight $\mu_e = 2$ for the interior of a white dwarf. A definition of the mean molecular weight and some examples are given in the chapter on the stellar atmospheres.

The two constant factors in (3) and (5) are computed in centimeter-gram-second units. The function $f(x)$ is a strictly increasing function of x, starting at $f(0) = 0$, with each positive value of x corresponding to a unique value $f(x)$ and vice-versa. In this way the density and the pressure are linked to each other in an unambiguous way. Therefore, it is important to realize that the equation of state does not contain the temperature as the ideal gas law does. This fact makes the computation of the structure of a white dwarf much easier than the structure of a normal main-sequence star. It is possible to compute the variations of density, pressure, radius and mass throughout the star without any knowledge of the temperature. The separation of the hydrostatic variables and the thermodynamic variables (T, energy, ...) simplifies the calculations considerably.

10.3 Equations of Structure

10.3.1 Equation of State

The equation of state for a white dwarf is the one for electron degeneracy and is given by (3,4,5). Given the density, it is easy to compute the value of x by (5) and then P by (3,4). By contrast, computing the density from the pressure requires some basic numerical techniques that will be explained in section 4.

10.3.2 Hydrostatic Equilibrium and Continuity of Mass

The equation of state is a local equation applicable at local points in the star. The two other relevant equations are differential equations that describe the variation of the pressure and the mass throughout the star as functions of the distance r to the center of the star. As usual M_r stands for the amount of mass inside a sphere with radius r. The equation of hydrostatic equilibrium containing the variation of the pressure is

$$dP = -\rho G \frac{M_r}{r^2} dr, \tag{6}$$

with G being the gravitational constant. A proof of this equation is given in the chapter on polytropes. The equation of continuity, giving the variation of mass as a function of the distance r is

$$dM_r = 4\pi \rho r^2 dr. \tag{7}$$

The goal of the next section is to solve these two equations in combination with the equation of state to obtain the structure of the white dwarf.

10.4 Numerical Method

10.4.1 Initial Conditions

The set of equations for the structure of white dwarfs contains two differential equations of the first order. This means that there are two initial conditions to be chosen. The first initial condition, the one for the mass as a function of the distance to the center is trivial: $M_r = 0$ at the center where $r = 0$. The second initial condition is the central pressure P_c. It is better, however, to start from a choice of the central density and use this to compute the central pressure with (3,4,5). For instance, taking a central density of 10^8 results in (5) having a value of the parameter $x = 3.7064$, and hence

$$f(X) = 354.307 \text{ and} \log P_c = 25.3282.$$

Since the initial condition for the mass is always zero, there is only one free initial condition for our problem. This single value will determine the complete structure of the white dwarf, its total mass, and its radius. We therefore say that the models for white dwarfs computed with this method are a one-parameter family of the central density. The choice of this central density is also limited since we are dealing with electron degeneracy. A lower limit for the logarithm of the central density is $\log \rho_c = 6$, while on the other hand we may take 12 as an upper limit. At higher densities the atomic nuclei also become degenerate, and we end up with a neutron star, once again with another equation of state. Therefore, the choice of the central density is limited by the condition

$$10^6 < \rho_c < 10^{12}. \tag{8}$$

10.4.2 Computing the Density from the Pressure

Before writing down the numerical formulae of the equations of structure, we will present a way to compute the density from the pressure. It is not difficult to compute the pressure from the density since in that direction we only have to compute x with (5) and introduce x in (3,4). On the other hand, the computation of ρ from P needs the computation of x for a given value of $f(x)$. For a given P we should compute x from

$$f(x) = \frac{P}{6.01 \times 10^{22}}.$$

Let us therefore first compute

$$K = \frac{P}{6.01 \times 10^{22}}. \tag{9}$$

The problem is then to find a value for x so that $f(x) = K$, or in other words we need to search the zero point of the function $f(x) - K$. This may easily be performed by successive approximations using Newton's method. This method is discussed in Chapter 1 on numerical techniques. In this case it may be applied in ideal circumstances since it is always possible to find a good starting value, and $f(x) - K$ is strictly monotone. The successive approximations are

$$x_{j+1} = x_j - \frac{f(x_j) - K}{f'(x_j)} \tag{10}$$

with $f(x)$ given by (4) and its derivative $f'(x)$ by

$$f'(x) = \frac{8x^4}{\sqrt{1 + x^2}}. \tag{11}$$

As a starting value we may use the x-value of the previous layer. Usually, less than five iterations are needed to obtain x with sufficient accuracy. In the author's program the iteration (10,11) is stopped when two subsequent values of x are found that differ by less than 0.000001. The criterion to stop the iteration is therefore

$$\mathrm{abs}(x_{j+1} - x_j) < 10^{-6}. \tag{12}$$

The last approximation is then used to compute the density with (5).

It is obvious that this procedure of finding x from $f(x)$ should be placed in a subprogram since we will need it two times during each step. A scheme for such a subprogram is presented in the following flowchart.

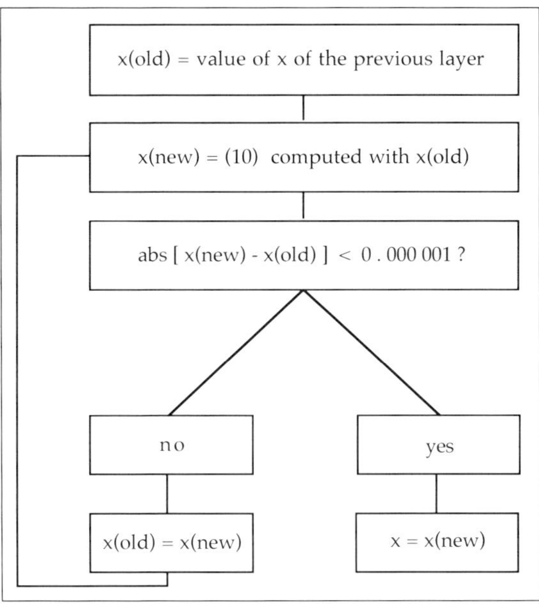

- *Example:* Compute the density from a pressure P, with $P = 2.5 \times 10^{25}$

  We find $f(x) = 415.9734$.

  Starting from $x = 5$, for instance, the approximations are as follows:

$$\begin{aligned} x(1) &= 4.1951366 \\ x(2) &= 3.8940479 \\ x(3) &= 3.8545814 \\ x(4) &= 3.8539539 \\ x(5) &= 3.8539537 \end{aligned}$$

With this value of x we find $\rho = 1.124245 \times 10^8$.

10.4.3 The Iteration Procedure

We will use the well-known midpoint method to calculate the white dwarf structure. The distance between two layers is given by dr which will be taken as constant throughout the star. The numerical formulae are

$$r_{i+1/2} = r_i + 0.5dr \tag{13}$$

$$P_{i+1/2} = P_i - 0.5dr\rho_i G\frac{M_{r,i}}{r_i^2} \tag{14}$$

$$M_{r,i+1/2} = M_{r,i} + 0.5dr4\pi\rho_i r_i^2 \tag{15}$$

$$x_{i+1/2} \text{ using } (10,11) \text{ with } K = \frac{P_{i+1/2}}{6.01 \times 10^{22}} \text{ starting at } x_i$$

$$\rho_{i+1/2} = 1.964 \times 10^6 x_{i+1/2}^3 \tag{16}$$

and then finally

$$r_{i+1} = r_i + dr \tag{17}$$

$$P_{i+1} = P_i - dr\rho_{i+1/2}G\frac{M_{r,i+1/2}}{r_{i+1/2}^2} \tag{18}$$

$$M_{r,i+1} = M_{r,i} + dr4\pi\rho_{i+1/2}\, r_{i+1/2}^2 \tag{19}$$

$$x_{i+1} \text{ using } (10,11) \text{ with } K = \frac{P_{i+1}}{6.01 \times 10^{22}} \text{ starting at } x_{i+1/2}$$

$$\rho_{i+1} = 1.964 \times 10^6 x_{i+1}^3. \tag{20}$$

Because of the presence of the radius in the denominator of equation (14), this equation may not be applied to the very first step when we proceed from the center ($r = 0$) to layer one. In that case we would divide by zero. We therefore need an alternative approach to the first step.

10.4.4 The First Step

At the first layer $r = dr$. The pressure and the mass may then be obtained by means of the Taylor series developed around zero:

$$P_1 = P_c - \frac{2}{3}G\pi\rho_c^2 dr^2 \tag{21}$$

$$M_{r,1} = \frac{4}{3}\pi\rho_c dr^3. \tag{22}$$

The density at layer one is then computed by (10,11) starting from x_0.

From this level on, we proceed with the iteration formulae given in section 4.3. The values at level zero (the center) are as follows:

$$\begin{aligned} r_0 &= 0 \\ M_{r,0} &= 0 \\ \rho_0 &= \text{free initial condition between } 10^6 \text{ and } 10^{12} \\ x_0 &= \text{from (5) with the central density} \\ P_0 &= \text{from } (3,4) \text{ with } x_0 \end{aligned}$$

We have used here the subscript 0 for the central values. They are the same as the values with subscript c.

The iteration procedure is stopped when a pressure turns out to be negative. This may occur at a certain level $i+1$ but also at a half-step, at $i+1/2$. Therefore, it *must* also be checked at half the step. The values of the radius and the mass at that moment are considered as the total radius and the total mass of the white dwarf. The small error we make by neglecting the fraction of the final unfinished step $i+1$ is never larger than about 2% of the total radius (if a normal step-size is used) and quasi zero for the mass. Professional programs sometimes contain instructions to compute the atmosphere of a white dwarf. This computation, however, would complicate our own program enormously. We have therefore limited the input physics to the model described above.

10.5 Some Remarks

The step-size dr should be given in centimeters and should be "small" compared to the solution area, in our case the radius of the star. We may hence select a step of a few times 10 km to 100 km. The Table 10-1 gives some possible step-sizes for various values of the central density. When you use these steps, the surface will be reached in about fifty steps. The values in the table are given in kilometers, don't forget to multiply by 100,000.

Table 10-1
The step-size (in km)
as a function of the central density

$\log \rho_c$:	5	6	7	8	9	10	11
dr(km) :	300	200	150	80	50	25	12

The equation of state used in the interior of white dwarfs is not valid in the outer shell of the star. In that region the effect of the temperature should be included. However this outer shell contains less than one percent of the radius and has no effect on the radius and the mass computed with our model. More information on the atmosphere of a white dwarf may be found in advanced textbooks on stellar structure.

The output data of the model are given in centimeter-gram-second units. As usual in astrophysics, we will list the logarithms of the density and the pressure, and express the masses in solar mass ($1M\odot = 2 \times 10^{33}$ g) and the distance and radius in solar radii ($1R\odot = 6.96 \times 10^{10}$ cm). The momentum parameter x is dimensionless.

10.6 Examples and Applications

10.6.1 Detailed Model of a White Dwarf

As a first test for your program, we present in Table 10-1 the complete results for a white dwarf characterized by a central density of 10^8 g/cm^3. The step-size used is 80 km or, in other words, 8,000,000 cm.

After layer 54 the pressure becomes negative, meaning that we have arrived at the surface of the white dwarf. The values of r and M_r are therefore considered as the total radius and mass of the white dwarf. We find for this model a radius of 0.0062069 solar radii, which is 4322.2 km (less than the Earth's radius), and a mass of 1.16 solar mass (375,000 times the mass of the Earth). This clearly demonstrates that white dwarfs are very dense objects. Their mass is comparable to the mass of a small star, but the radius is only that of a small planet. The average density is more than 100,000 times the average density of the Earth and more than 400,000 times the average density of the Sun. Don't worry if the results differ in the last few significant digits. This discrepancy is a consequence of rounding effects.

10.6.2 The Chandrasekhar Limit

The only free parameter in our model is the central density. It determines the whole structure, the mass and the radius. The dependency of these quantities as a function of the central density has been studied by the Indian astronomer S. Chandrasekhar, one of the greatest pioneers in the study of stellar structure. His conclusions are twofold:

- The radius decreases and the mass increases as a function of the central density, indicating that the smallest white dwarfs also have the largest mass.

- There is an upper limit for the mass of white dwarfs.

This upper limit, the "Chandrasekhar limit," is the largest mass a white dwarf can have: $1.44M\odot$. It is reached when the central density gets larger

Table 10-2

i	r	M_r	$\log P$	$\log \rho$	x
0	0.0000000	0.0000000	25.3282545	8.0000000	3.7064047
1	0.0001149	0.0001072	25.3264264	7.9986669	3.7026143
2	0.0002299	0.0008266	25.3223687	7.9957082	3.6942157
3	0.0003448	0.0028057	25.3143447	7.9898585	3.6776664
4	0.0004598	0.0066209	25.3024675	7.9812019	3.6533124
5	0.0005747	0.0127843	25.2868627	7.9698326	3.6215712
6	0.0006897	0.0217263	25.2676438	7.9558366	3.5828753
7	0.0008046	0.0337837	25.2449163	7.9392949	3.5376739
8	0.0009195	0.0491918	25.2187876	7.9202904	3.4864463
9	0.0010345	0.0680817	25.1893685	7.8989093	3.4296985
10	0.0011494	0.0904827	25.1567734	7.8752410	3.3679567
11	0.0012644	0.1163279	25.1211198	7.8493776	3.3017589
12	0.0013793	0.1454634	25.0825269	7.8214131	3.2316467
13	0.0014943	0.1776604	25.0411147	7.7914425	3.1581567
14	0.0016092	0.2126283	24.9970016	7.7595604	3.0818132
15	0.0017241	0.2500288	24.9503038	7.7258600	3.0031211
16	0.0018391	0.2894899	24.9011332	7.6904319	2.9225603
17	0.0019540	0.3306195	24.8495963	7.6533634	2.8405817
18	0.0020690	0.3730174	24.7957930	7.6147370	2.7576036
19	0.0021839	0.4162861	24.7398150	7.5746303	2.6740095
20	0.0022989	0.4600402	24.6817453	7.5331146	2.5901468
21	0.0024138	0.5039135	24.6216567	7.4902546	2.5063268
22	0.0025287	0.5475650	24.5596112	7.4461074	2.4228246
23	0.0026437	0.5906827	24.4956590	7.4007223	2.3398802
24	0.0027586	0.6329864	24.4298373	7.3541401	2.2576998
25	0.0028736	0.6742292	24.3621702	7.3063922	2.1764577
26	0.0029885	0.7141974	24.2926666	7.2575005	2.0962979
27	0.0031034	0.7527106	24.2213202	7.2074765	2.0173364
28	0.0032184	0.7896198	24.1481075	7.1563204	1.9396631
29	0.0033333	0.8248060	24.0729866	7.1040206	1.8633440
30	0.0034483	0.8581783	23.9958955	7.0505524	1.7884232
31	0.0035632	0.8896711	23.9167498	6.9958766	1.7149247
32	0.0036782	0.9192421	23.8354397	6.9399383	1.6428539
33	0.0037931	0.9468698	23.7518273	6.8826646	1.5721996
34	0.0039080	0.9725511	23.6657418	6.8239619	1.5029348
35	0.0040230	0.9962988	23.5769743	6.7637134	1.4350179
36	0.0041379	1.0181399	23.4852705	6.7017741	1.3683931
37	0.0042529	1.0381133	23.3903223	6.6379660	1.3029912
38	0.0043678	1.0562680	23.2917557	6.5720709	1.2387294
39	0.0044828	1.0726616	23.1891149	6.5038211	1.1755105
40	0.0045977	1.0873590	23.0818417	6.4328868	1.1132220
41	0.0047126	1.1004310	22.9692469	6.3588589	1.0517338
42	0.0048276	1.1119531	22.8504695	6.2812254	0.9908958
43	0.0049425	1.1220052	22.7244207	6.1993366	0.9305329
44	0.0050575	1.1306702	22.5897003	6.1123562	0.8704388
45	0.0051724	1.1380341	22.4444720	6.0191861	0.8103666
46	0.0052874	1.1441854	22.2862700	5.9183509	0.7500147
47	0.0054023	1.1492150	22.1116878	5.8078115	0.6890065
48	0.0055172	1.1532166	21.9158552	5.6846527	0.6268598
49	0.0056322	1.1562869	21.6915228	5.5445363	0.5629437
50	0.0057471	1.1585265	21.4274073	5.3807141	0.4964294
51	0.0058621	1.1600418	21.1053432	5.1823359	0.4263159
52	0.0059770	1.1609479	20.6982796	4.9333119	0.3521466
53	0.0060920	1.1613732	20.2095516	4.6361666	0.2803335
54	0.0062069	1.1614592	19.9948103	4.5060701	0.2636936

than 10^{12} g/cm^3. Higher masses are not possible for white dwarfs. In that case the star becomes a neutron star, to be described by physical laws other than those we have used here. With your program you are able to check the value of the Chandrasekhar limit by computing a series of models with increasing central density. Table 10-3 lists our results. Some values of the central density may look strange, but they were selected to compare our results with professional data. The latter are listed between parentheses. The professional results are based on Chandrasekhar's hydrogen degenerate configurations[1] which we have converted to pure helium with (23,24).

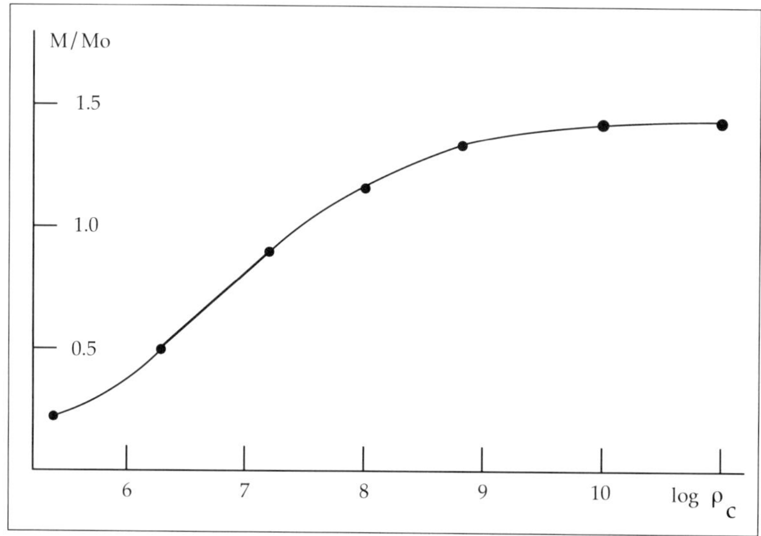

Fig. 10.3: *The mass versus the central density computed with our program.*

Table 10-3
Masses and radii computed with our program
compared with the converted professional results
(in parentheses).

$\log \rho_c$	M/M_0	$M/M_\odot$ $\log R/R_\odot$
5.39	0.22 (0.22)	−1.70 (−1.70)
6.29	0.50 (0.50)	−1.86 (−1.86)
7.20	0.89 (0.88)	−2.03 (−2.03)
8.00	1.16	−2.21
8.83	1.33 (1.33)	−2.41 (−2.41)
10.00	1.42	−2.72
11.00	1.435	−3.02

[1] *An Introduction to the Study of Stellar Structure,* 1939, University of Chicago Press.

The mass depends in a quasi-linear way on the logarithm of the central density for densities below 10^8, but bends to a constant value of 1.44 solar mass for higher densities.

We have adopted in all the equations and parameters in this chapter the mean molecular weight $\mu_e = 2$. This means that our white dwarfs consist entirely of pure helium. The mean molecular weight will have another value when a different chemical composition is assumed. In that case it is not necessary to recompute the whole structure of the white dwarf. Considering two white dwarfs A and B, with different chemical composition, their physical variables are related by the following equations:

$$r(A)\mu_e(A) = r(B)\mu_e(B) \quad \rho(A)\mu_e(A) = \rho(B)\mu_e(B) \tag{23}$$

$$P(A) = P(B) \quad M_r(A)\mu_e(A)^2 = M_r(B)\mu_e(B)^2. \tag{24}$$

These formulae should be applied separately at every layer.

10.7 The Program Listing

The following sample program is written in MicroSoft QuickBasic 4.5:

```
DECLARE FUNCTION press (m, r, d)
DECLARE FUNCTION f (x)
DECLARE FUNCTION lg10 (x)
DECLARE FUNCTION dfdx (x)
DECLARE SUB density (p, xold, rho, xnew)
DECLARE FUNCTION pess (m, r, d)
DECLARE FUNCTION mass (d, r)


COMMON SHARED g, m0, r0, a, pi

' some physical constants used in this program :
g = 6.673E-08
m0 = 2E+33
r0 = 6.96E+10
a = 6.01E+22
pi = 3.1415926536#


CLS
PRINT "Astrophysics With a PC : WHITE DWARF"
PRINT "----------------------------------------"
PRINT " "
PRINT "-------- Minimal solution program --------"
PRINT " "
PRINT "Input of initial conditions and model parameters : "
INPUT "Log10 of the central density : ", rhc
INPUT "stepsize in km                : ", drkm

'central values of the physical variables
rhoc = 10 ^ rhc
dr = 100000! * drkm
PRINT " "
m = 0!
r = 0!
xc = (rhoc / 1964000!) ^ (1 / 3)
```

```
pc = a * f(xc)
i% = 0

' display heading of the table with the results
PRINT "  i     r          Mr        log(P)    log(rho)      x"

' display central values (layer zero)
PRINT USING "###  ###.#####  ###.#####  ###.#####  ###.#####  ##.#######";_
 i%; r / r0; m / m0; lg10(pc); lg10(rhoc); xc

' compute first step from layer zero to layer one, and show them on screen
p = pc - 2 / 3 * g * pi * (rhoc * dr) ^ 2
m = 4 / 3 * pi * rhoc * dr ^ 3
CALL density(p, xc, d, x)
r = dr
i% = 1
PRINT USING "###  ###.#####  ###.#####  ###.#####  ###.#####  ##.#######";_
 i%; r / r0; m / m0; lg10(p); lg10(d); x

' prepare for the other layers (from layer two to the surface
surface% = 0
i% = 2

' here starts the main cycle for the other layers
DO

    r1 = r + .5 * dr
    p1 = p + .5 * dr * press(m, r, d)

' check if pressure (half step i+1/2) is still positive,
' otherwise surface is reached
    IF p1 > 0
    THEN
        m1 = m + .5 * dr * mass(d, r)
        CALL density(p1, x, d1, x1)
        r = r + dr
        p = p + dr * press(m1, r1, d1)

'      check if pressure (full step i+1) is still positive,
'      otherwise surface is reached
        IF p > 0
        THEN
            m = m + dr * mass(d1, r1)
            CALL density(p, x1, d, x)

'          show results of newly computed layer on screen
            PRINT USING "###  ###.#####  ###.#####  ###.#####  ###.#####_
  ##.#######"; i%; r / r0; m / m0; lg10(p); lg10(d); x
            i% = i% + 1

'          pause after every 12 layers
            IF i% MOD 12 = 0
            THEN
                PRINT "Press any key to continue"
                DO
                LOOP WHILE INKEY$ = ""
            END IF
        ELSE                 'this ELSE is reached if pressure (i+1)_
 was negative
            surface% = 1
        END IF
```

```
   ELSE                      'this ELSE is reached if pressure (i+1/2)_
was negative
       surface% = 1
   END IF
LOOP UNTIL surface% = 1


' show general results on screen
PRINT " "
PRINT USING "Total mass : ##.##"; m / m0


PRINT "Program terminated. Press any key to stop"
DO
LOOP WHILE INKEY$ = ""
END



SUB density (p, xold, rho, xnew)
'
' computes the density rho and the corresponding value xnew from the
' input pressure P and starting guess xold
'
     k = p / a
     stopcrit% = 0
     DO
           xnew = xold - (f(xold) - k) / dfdx(xold)
           IF ABS(xold - xnew) < .000001
           THEN
               stopcrit% = 1
           ELSE
               xold = xnew
           END IF
     LOOP UNTIL stopcrit% = 1
     rho = 1964000! * xnew ^ 3
END SUB



FUNCTION dfdx (x)
'
' evaluates the derivative of f(x)
'
     dfdx = 8 * x ^ 4 / SQR(1 + x ^ 2)
END FUNCTION



FUNCTION f (x)
'
' evaluates the function f(x)
'
     f = x * (2 * x ^ 2 - 3) * SQR(1 + x ^ 2) + 3 * LOG(x + SQR(1 + x ^ 2))
END FUNCTION



FUNCTION lg10 (x)
'
' computes the base 10 logarithm of x using the natural logarithm LOG(x)
'
    lg10 = LOG(x) / LOG(10!)
END FUNCTION



FUNCTION mass (d, r)
'
' computes the right hand side of the differential equation of mass continuity
```

```
'
    mass = 4 * pi * d * r * r
END FUNCTION


FUNCTION press (m, r, d)
'
' computes the right hand side of the equation of hydrostatic equilibrium
'
    press = -g * m * d / r ^ 2
END FUNCTION
```

Chapter 11

Star Formation in the Galaxy

11.1 Introduction

Blue stars are the most luminous. Located in the upper left part on the main sequence in the Hertzsprung-Russell diagram, these stars are also the most massive of all the main-sequence stars and have the largest radius. Their larger mass, however, is not able to compensate for the high luminosity. This means that, although their supply of hydrogen is larger than stars like the Sun, they will burn up this supply at such a high rate that their total lifetime is much shorter than for medium or low mass stars.

Why is it important to look for these short-living stars? Since their total lifetime is so short, they will not have had the opportunity to move away very far from the place where they were born. Consequently, if we want to search for galactic regions in which star formation is possible, we have to look for places with luminous blue stars. This is not always possible in our own galaxy where our view is blocked by interstellar clouds, but there are more than enough other galaxies. Luminous blue stars are found in the outer regions of spiral galaxies. The spiral components of these galaxies actually have a blue color because the luminosity of these blue stars is so high that it compensates for their small number. The spiral arms also have another important component, namely extended hydrogen and dust clouds, whose presence leads to the suggestion that stars are born in these clouds by condensation and gravitational collapse. A number of pictures taken from dark clouds in the galaxy clearly show spherical condensations, called *globules* or *protostars.* Young stars are often found in the immediate surroundings of these protostars.

A normal consequence of contraction of a gas cloud is an increase of its temperature. During the first stages, however, the radiation is able to escape so that both the temperature and the pressure in the cloud stay at a low level. The cloud will therefore continue to collapse, eventually fragmenting into a number of smaller contracting clouds. By the time the central density becomes high enough to make the center opaque for infrared radiation, the gravitational field is strong enough to compensate for the increasing pressure. The collapse is now inevitable and will only come to an end when the central temperature reaches several million K. This is the minimum temperature to start the nuclear fusion of hydrogen into helium, a process that releases an enormous amount of energy. This energy is transported through the cloud and radiated away in space. The flow of energy also restores hydrostatic equilibrium. The cloud has now become a normal core hydrogen burning star.

An important amount of kinetic energy is released during the contraction stage. It is necessary that this energy be neutralized without heating the cloud to prevent any further contraction. CO molecules play an especially important role as cooling mechanisms in dust clouds. We present here a first approximation model from G. Bodifée (Ph.D thesis, Free University of Brussels, 1985). A detailed description of the physics involved and an analytical treatment of the stability is given by Bodifée (*Astrophysics and Space Science* **122**, 41, 1986). Our approach is completely different since we solve the equations numerically. A more complicated model with more physical processes included was presented by Bodifée and De Loore (*Astron. Astrophys.* **142**, 297, 1985).

11.2 A Simple Model for Star Formation

The model contains three active components: cool atomic clouds, cool molecular clouds, and active young stars. Each of these components may interact with the other components or with the rest of the galaxy. The model is therefore an open system connected with two mass reservoirs outside the system. A schematic view of the system is presented in Fig. 11.1.

The main component of cool atomic clouds is neutral hydrogen, the most abundant chemical element in the galaxy. Neutral hydrogen clouds are found in the spiral arms and the disk of the galaxy. Their typical radio emission at a wavelength of 21 cm enabled Van de Hulst to construct a map of the spiral structure of the galaxy. The density varies over a large range, but direct star formation in these clouds seems not to occur. The cooling capacity of these clouds is not large enough to allow a sufficient condensation. This component is connected to an unlimited reservoir of new atomic gas outside the system.

Cool molecular clouds mainly consist of molecular hydrogen HII. The densities may vary enormously and are generally much higher than in neutral clouds. The temperature in such a cloud decreases as its density increases as a consequence of the large cooling capacity of the CO molecules. Molecu-

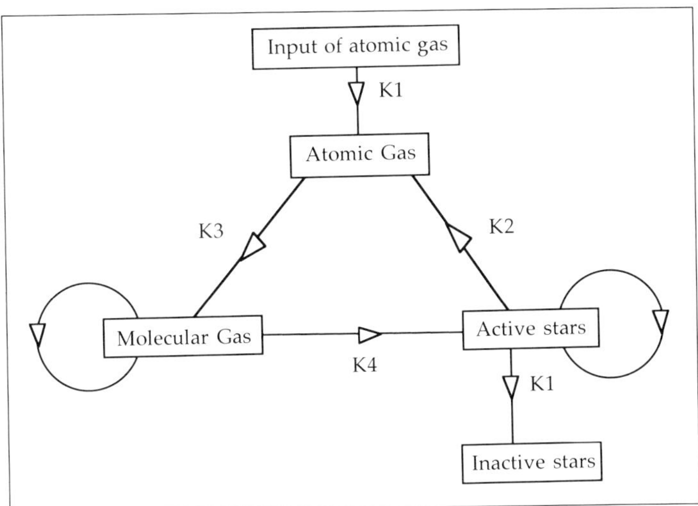

Fig. 11.1: *A schematic view of the various components of the star formation model of Bodifée.*

lar clouds have smaller dimensions than neutral clouds. They are found in the spiral arms, often in the vicinity of OB associations, which are groups of young blue stars.

Young, active stars are mostly accompanied by hot ionized HII gas. These stars strongly affect the surrounding gas clouds and are responsible for shock waves in these clouds. In this way, new condensation regions may be formed in the molecular clouds of the system. The presence of young stars therefore has a positive effect on the stellar birth rate. The capacity of influencing the other components ends when these young stars evolve to neutron stars. Although these remnants are still physically present in the system, their mass has stopped playing an active role in the star formation process. We therefore say that this mass has left the active star formation system. The second reservoir is hence a waste reservoir containing the stellar remnants.

11.3 Equations of Interaction Between the Components

The three mass components will be described by the variables S for the total mass of active stars, M for the total mass of molecular clouds, and A for the total mass of atomic clouds. It is assumed that the total mass of the system remains constant; thus, we assume that the amount of mass lost by stellar evolution is exactly replaced by fresh atomic clouds entering the star formation region from the rest of the galaxy. If we call the total mass of the system T, we may write

$$T = A + M + S. \tag{1}$$

There are three kinds of interaction for the atomic cloud component A. First, there is a constant replenishment by new atomic clouds in an amount equal to the amount of mass leaving the active system by stellar evolution. The amount of new gas may therefore be considered as proportional to the amount of stellar mass S. We will call the proportional constant of this process K_1. Secondly, the atomic component is increased as young, active stars lose mass by stellar wind. This process is also proportional to the number of stars and therefore to the amount of stellar mass. The proportional constant for this process is K_2. The third interaction is the transformation of atomic into molecular clouds. This process is clearly proportional to the amount of atomic gas A, but since this transformation becomes more and more effective with the cooling capacity of the cloud, and since this capacity increases with the square of the density of the molecular content, we also assume that the transformation of atomic into molecular gas is proportional to the square of the molecular mass. This third process is a loss of atomic gas and therefore is written with a minus sign in the differential equation and a proportional constant K_3:

$$\frac{dA}{dt} = K_1 S + K_2 S - K_3 A M^2. \tag{2a}$$

According to Schmidt (*Astrophys. J.*, **129**, 243, 1959), the rate of star formation may be considered as being proportional to the n-th power of the density of the molecular cloud. Values of n can be selected between 0.5 and 3.5 or, to play it safe, between 1 and 2. It is assumed that the presence of other young stars is a necessary condition for star formation since they will perturb the molecular cloud and provoke condensations. In this way we may also state that the star formation rate is proportional to the number of active stars already present. Let us call the proportional constant K_4. This process increases the mass of the stellar component. Two other processes decrease it: stellar evolution, for which we may use K_1, and mass loss by stellar wind, for which we again use K_2. Both processes are proportional to the amount of stellar mass. Thus, the equation describing the variation of the stellar mass in the system is

$$\frac{dS}{dt} = K_4 S M^n - K_1 S - K_2 S. \tag{2b}$$

Finally, the variation of the total molecular mass is given by two processes already described: transformation of atomic into molecular gas, which is an increase for the variable M, and stellar formation, which decreases the amount of molecular mass. The equation for M is

$$\frac{dM}{dt} = K_3 A M^2 - K_4 S M^n. \tag{2c}$$

This completes the set of equations. It is easy to see that every process in one of the three differential equations is compensated by the same process with an opposite sign in one of the other equations. This balance guarantees the conservation of the total amount of mass.

11.4 Numerical Methods

Bodifée has simplified the solution of his model by eliminating one equation and by using dimensionless variables. The three independent variables A, M, and S in the previous section are linked by the fact that we consider the total mass T of the star formation system as constant. One may, for instance, replace S by T–A–M in equations (2a) and (2c). The system then reduces to only two first-order equations. Furthermore, it is possible to transform these two remaining equations by introducing new dimensionless variables and a new dimensionless time coordinate x.

$$a = \frac{A}{T} \tag{3a}$$

$$m = \frac{M}{T} \tag{3b}$$

$$s = \frac{S}{T} \tag{3c}$$

$$x = (K_1 + K_2)t \tag{3d}$$

This implies that

$$a + m + s = 1 \tag{4}$$

at every moment. The parameters K_1, K_2, K_3, and K_4 are also transformed as a consequence of the dimensionless variables and become two new parameters:

$$k_1 = \frac{K_3 T^2}{K_1 + K_2} \tag{5a}$$

and

$$k_2 = \frac{K_4 T^n}{K_1 + K_2}. \tag{5b}$$

When two out of the three variables are known, the third also is known since the sum of the three is always 1. The differential equations, after elimination of S and after introducing the new variables, become

$$\frac{da}{dx} = 1 - a - m - k_1 m^2 a \tag{6a}$$

and

$$\frac{dm}{dx} = k_1 m^2 a + k_2 m^n (a - 1 + m). \tag{6b}$$

These are the differential equations to be solved. They contain three free parameters: k_1, k_2, and n. As usual, we'll apply a midpoint method for these two coupled first-order equations. The numerical formulae for a time step dx are given below:

$$x_{i+1/2} = x_i + \frac{1}{2}dx \tag{7a}$$

$$a_{i+1/2} = a_i + \frac{1}{2}dx\Big(1 - a_i - m_i - k_1 m_i^2 a_i\Big) \tag{7b}$$

$$m_{i+1/2} = m_i + \frac{1}{2}dx\Big(k_1 m_i^2 a_i + k_2 m_i^n (a_i - 1 + m_i)\Big) \tag{7c}$$

$$x_{i+1} = x_i + dx \tag{8a}$$

$$a_{i+1} = a_i + dx\Big(1 - a_{i+1/2} - m_{i+1/2} - k_1 m_{i+1/2}^2 a_{i+1/2}\Big) \tag{8b}$$

$$m_{i+1} = m_i + dx\Big(k_1 m_{i+1/2}^2 a_{i+1/2} + k_2 m_{i+1/2}^n (a_{i+1/2} - 1 + m_{i+1/2})\Big) \tag{8c}$$

$$s_{i+1} = 1 - m_{i+1} - a_{i+1}. \tag{9}$$

Time steps dx of a few times 0.01 are stable. All our examples are computed with a time step $dx = 0.02$.

To obtain meaningful results, typical values for the three free parameters should be selected in the following way:

n should be a real number between 1 and 2,

k_1 should be a positive real number up to 100,

k_2 should be a positive real number up to 100.

The initial values of the two differential equations should be selected according to the fact that the total sum of the three mass fractions is 1. Hence, the initial conditions a_0 and m_0 should both be taken between zero and one under the supplementary condition that their sum is smaller than or equal to one. If their sum were larger, there would be a negative stellar component—which is, of course, impossible. The initial stellar fraction is easily obtained from

$$s_0 = 1 - m_0 - a_0.$$

The initial stellar fraction s_0 is then also between zero and one. It is important to note that every term in the right-hand side of the equation for star formation (2b) is proportional to the stellar fraction S. This means that if S equals zero at the start, it will stay zero. No stars will be formed in that case. The only process which will occur is a transformation of atomic gas into molecular gas. However, if you actually try this, you will see that after all the gas has become molecular, stars will be formed as a consequence of rounding errors in the computer program!

11.5 Applications

First, we present some numerical results of a model with the following parameter values and initial conditions:

$$\begin{aligned} n &= 1 \\ k_1 &= 10 \\ k_2 &= 2.5 \\ m_0 &= 0.15 \\ s_0 &= 0.75 \\ a_0 &= 0.1 \end{aligned}$$

The iterations are computed with a time step $dx = 0.02$.

i	x	a	m	s
0	0.00	0.1000	0.1500	0.7500
1	0.02	0.1144	0.1450	0.7406
2	0.04	0.1287	0.1402	0.7311
3	0.06	0.1427	0.1357	0.7216
4	0.08	0.1565	0.1315	0.7120
5	0.10	0.1701	0.1274	0.7025
6	0.12	0.1835	0.1236	0.6929
7	0.14	0.1967	0.1200	0.6833
8	0.16	0.2097	0.1166	0.6738
9	0.18	0.2225	0.1133	0.6642
10	0.20	0.2351	0.1102	0.6547
11	0.22	0.2475	0.1072	0.6453
12	0.24	0.2598	0.1044	0.6358
13	0.26	0.2718	0.1017	0.6265
14	0.28	0.2837	0.0991	0.6171
15	0.30	0.2954	0.0967	0.6079
16	0.32	0.3069	0.0944	0.5987
17	0.34	0.3183	0.0921	0.5896
18	0.36	0.3294	0.0900	0.5806
19	0.38	0.3404	0.0880	0.5716

These results enable you to check whether your program is correct.

The results may graphically be represented in two ways. It is, of course, possible to plot the three variables—the atomic, molecular, and stellar content—as functions of the time. This is done for a number of models whose results will be discussed later. Another possible method often used in all applied sciences is to work with phase diagrams. Consider, for instance, a mass hanging on a spring. The mass will jump up and down, periodically oscillating around the state of rest. We may once again plot the position and the velocity as functions of the time in two separate drawings, but another way is to plot the position X on the horizontal axis and the corresponding velocity V on the vertical axis. We will then obtain a point that is moving on a circle in the XV-plane. If the oscillations are damped, the radius will slowly decrease until the circle shrinks to a point on the X-axis. This means that the position is constant and the velocity is zero. The mass has come to rest.

Diagrams in which two functions of the time are plotted on the two axes are called *phase diagrams*, just as if the time had been eliminated from the two functions so that we have obtained one single relation containing the two variables. Phase diagrams are especially suited to the study of periodic or oscillating systems.

11.5.1 Evolution Towards a Stationary State

The model evolves towards a stationary state for certain combinations of the three free parameters. For fixed values of n and k_2, there is some critical value of k_1 above which the models evolve to such a state. A detailed analysis of these critical values was performed by Bodifée. The model first starts with some periodic oscillations. These oscillations are then damped, and after a certain relaxation time, all variables (i.e., the atomic, molecular, and stellar content) reach constant values. This means that all the interactions between the three components are still present, but in such a way that their combined effects cancel out. The decrease of the molecular fraction by star formation, for instance, is perfectly balanced by the increase caused by the transformation of atomic into molecular gas. The rate of star formation is also constant.

- *Example:*

  Parameters:

$$n = 1.0$$
$$k_1 = 10$$
$$k_2 = 2.5$$

  Initial conditions:

$$m_0 = 0.15$$
$$s_0 = 0.75$$
$$a_0 = 0.1$$

The results are presented in Figs. 11.2a and 11.2b. They contain 2000 iterations with a time step $dx = 0.02$. We start with a galactic subsystem with a high stellar mass fraction of 0.75. These stars all evolve together to an inactive state, meaning that their mass is regenerated in the system in the form of atomic gas. This is why the atomic content increases very sharply in

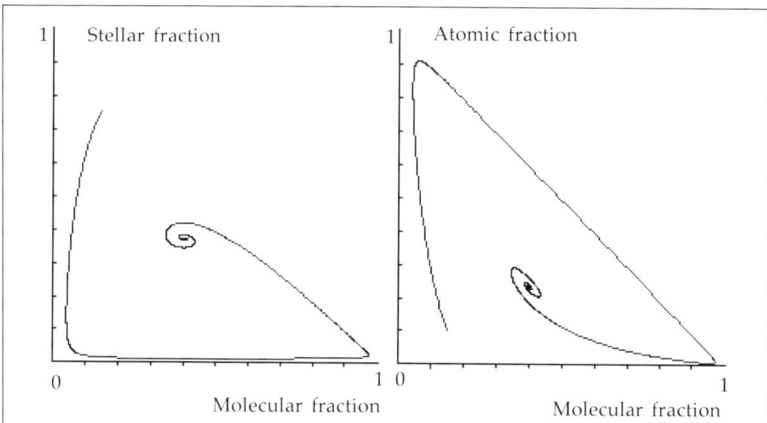

Fig. 11.2a: *Phase diagram of star formation model. The model evolves to a stationary state.* $n = 1.0$, $k_1 = 10$, $k_2 = 2.5$, $m_0 = 0.15$, $s_0 = 0.75$, $a_0 = 0.10$. *2000 iterations with* $dx = 0.02$.

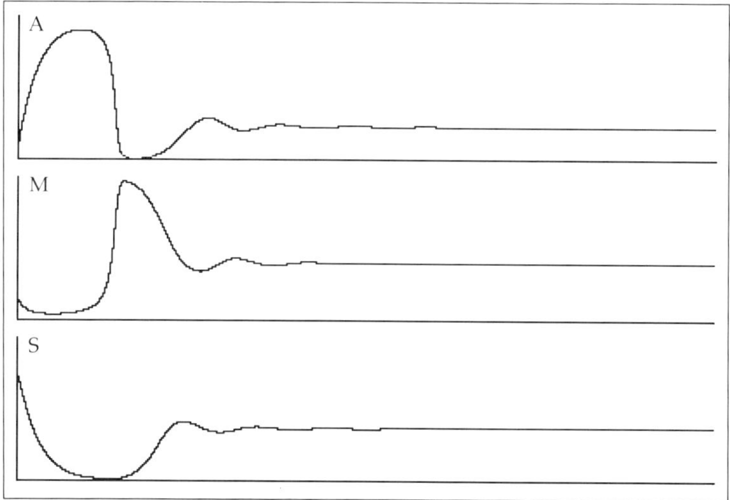

Fig. 11.2b: *Atomic, molecular, and stellar fractions as functions of the time; same initial conditions as Fig. 11.2a.*

the beginning (see Fig. 11.2a). All this atomic gas is then transformed completely into molecular gas, a process visible as the diagonal decrease on the right side of Fig. 11.12a. Finally, the system evolves after some very quickly damped oscillations towards a stationary state with mass fraction of about 0.40 for the molecular clouds, 0.35 for the stellar content, and 0.25 for the atomic clouds. These final and stable fractions are represented by the constant levels in Fig. 11.2b.

11.5.2 Evolution Towards a Limit Cycle

Another possible final state of the star formation model is the limit cycle. In this case, the model evolves into an oscillating but stable state. We present two examples:

- *Example:*

  Parameters:
  $$n = 1.7$$
  $$k_1 = 20$$
  $$k_2 = 25$$

  Initial conditions:
  $$m_0 = 0.7$$
  $$s_0 = 0.2$$
  $$a_0 = 0.1$$

- *Example:*

  Parameters:
  $$n = 1.5$$
  $$k_1 = 8$$
  $$k_2 = 15$$

  Initial conditions:
  $$m_0 = 0.3$$
  $$s_0 = 0.3$$
  $$a_0 = 0.4$$

The results are shown in Figs. 11.3 and 11.4, respectively. The period and amplitude of the oscillations are constant. In particular, this means for the stellar fraction that the birth rate is not constant in time. Star formation is concentrated at certain moments as is easily seen in the time dependency diagrams 3b and 4b. In the case of Fig. 11.3, there are always active stars in the system, at least about 5%. The oscillations of Fig. 11.4 are more violent and this model is characterized by large periods in which the major part of the mass is atomic gas, and nearly all the rest molecular. Almost no active stars are present. Then there is a quite sudden transformation of almost all the atomic gas into molecular gas, immediately followed by a burst of star formation. Then all these stars evolve more or less, leaving the active star formation system. Their mass is replaced by fresh atomic gas, which becomes again the most important component.

We are indebted to Dr. G. Bodifée for providing the model equations to us and for valuable discussions on their application.

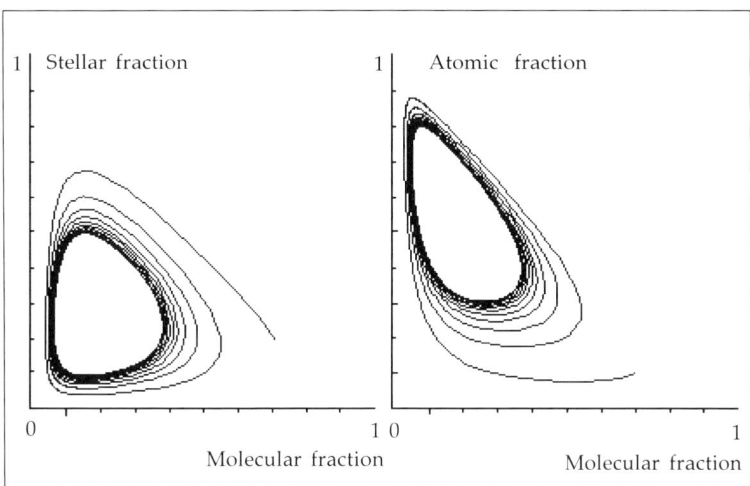

Fig. 11.3a: *Phase diagram of star formation model. The model evolves to a limit cycle.* $n = 1.7$, $k_1 = 20$, $k_2 = 25$, $m_0 = 0.70$, $s_0 = 0.20$, $a_0 = 0.10$. *2000 iterations with* $dx = 0.02$.

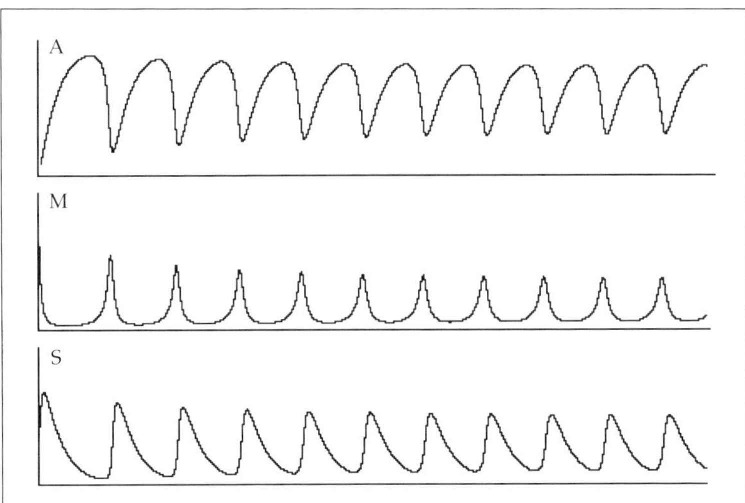

Fig. 11.3b: *Atomic, molecular, and stellar fractions as functions of the time; same initial conditions as Fig. 11.3a.*

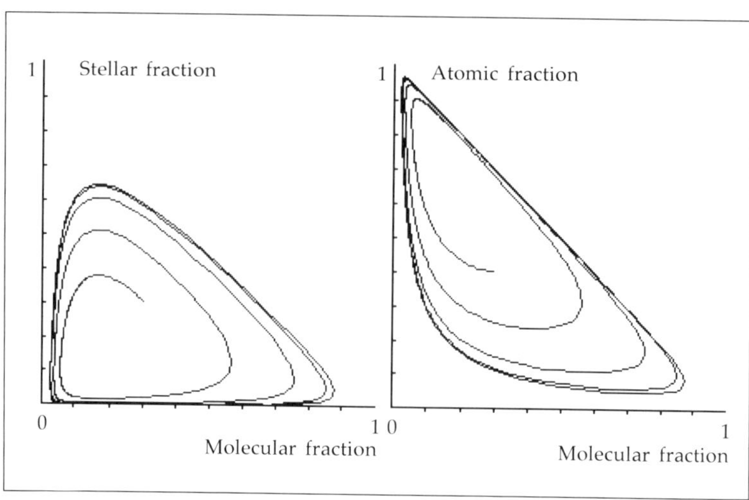

Fig. 11.4a: *Phase diagram of star formation model. The model evolves to a limit cycle.* $n = 1.5$, $k_1 = 8$, $k_2 = 15$, $m_0 = 0.3$, $s_0 = 0.3$, $a_0 = 0.4$. *2000 iterations with* $dx = 0.02$.

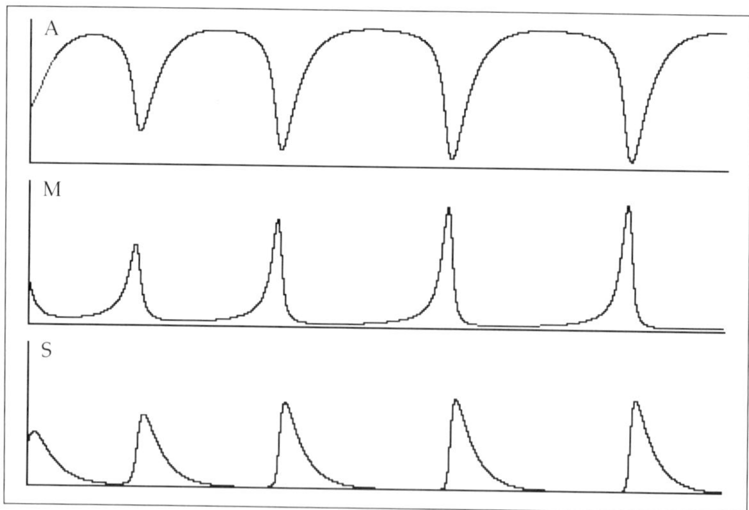

Fig. 11.4b: *Atomic, molecular, and stellar fractions as functions of the time; same initial conditions as Fig. 11.4a.*

11.6 The Program Listing

The following sample program is written in MicroSoft QuickBasic 4.5:

```
DECLARE SUB effes (a, m, k1, k2, n, fa, fm)


CLS
PRINT "Astrophysics With a PC : GALACTIC STARFORMATION"
PRINT "-------------------------------------------------"
PRINT " "
PRINT "------------ Minimal solution program -------------"
PRINT " "
PRINT "Input of initial conditions and parameters : "
INPUT "Initial fraction of molecular clouds : ", m
INPUT "Initial fraction of atomic gas : ", a

INPUT "Parameter   n : ", n
INPUT "Parameter k1  : ", k1
INPUT "Parameter k2  : ", k2
PRINT " "



s = 1 - a - m
dx = .02


' heading of main cycle
PRINT "    i        x          a              m              s"


' prepare for main cycle
x = 0
ni% = 1


' here start main cycle, computing blocks of 20 iterations
DO
    FOR i% = 1 TO 20


'      values at half the step (i+1/2)
        CALL effes(a, m, k1, k2, n, fa, fm)
        m1 = m + .5 * dx * fm
        a1 = a + .5 * dx * fa
        s1 = 1 - m1 - a1


'       values at the full step (i+1)
        CALL effes(a1, m1, k1, k2, n, fa, fm)
        m = m + dx * fm
        a = a + dx * fa
        s = 1 - m - a
        x = x + dx


'       show results of newly computed layer on screen
        PRINT USING " ####   #####.####   ###.#######   ###.#######_
   ###.#########"; ni%; x; a; m; s


        ni% = ni% + 1
    NEXT i%


'  Ask user whether to compute 20 iterations more or not
    PRINT "Press S to stop or any other key to continue"
    DO
        ch$ = INKEY$
        LOOP WHILE ch$ = ""


LOOP UNTIL (ch$ = "s") OR (ch$ = "S")
```

```
END


SUB effes (a, m, k1, k2, n, fa, fm)
'
' computes the right hand sides of the differential equations for the
' variables a (atomic fraction) and m (molecular fraction)
'
    fa = 1 - a - m - k1 * m ^ 2 * a
    IF m > 0!
    THEN
        fm = k1 * m ^ 2 * a + k2 * (a - 1 + m) * m ^ n
    ELSE
        fm = 0!
    END IF
END SUB
```

Chapter 12

Individual Stellar Orbits in the Galaxy

12.1 Introduction

Our galaxy contains about 100 billion stars. All these stars, combined with interstellar clouds of various nature and globular clusters present in the halo outside the galactic disc, both determine and suffer from the galactic gravitational field. All the stars take part in the general galactic rotation, but the orbits are also subject to radial displacements away from, and in the direction of the galactic center, and to oscillations above and under the galactic plane. Such individual motions may be observed from the solar system as small changes in the coordinates of the stars and as small blue or red Doppler shifts in their spectra. Doppler shifts of galactic stars correspond to velocities up to a few times 10 km/s. When measured in a fixed coordinate system, the individual velocities are higher since the tangential velocity due to the rotation of the galaxy as a whole may be up to 300 km/s.

Before focusing on orbits of individual stars, we first have to adopt a model for the galaxy itself. Various models are possible, from very simple to very complicated constructions. The most simple models—such as the point model (in which all the matter is concentrated in one central point) or the homogeneous spherical model—are very simple indeed from a mathematical point of view, but they are unable to reproduce the observed fundamental physical properties. We will adopt a model of the Schmidt type: the non-homogeneous oblate spheroid model, described by M. Schmidt in his Ph.D. thesis (1956). This model, with some modifications for our own application, gives us a suitable compromise between mathematical simplicity and physical reality.

Any model is defined by its mass distribution, i.e., the mass density at every point in the galaxy. This distribution determines the gravitational field, the rotational curve and the equations of motion for an individual object with negligible mass. It is assumed that the mass distribution is constant in time: the effects of the motions of the individual stars cancel each other out. The motion of one star may then be considered as taking place in a fixed gravitational field, depending only on the position. Furthermore, the motion of an individual star has no effect on the gravitational field. The results obtained from observations justify the assumption that the mass distribution is symmetric around the rotational axis (perpendicular on the disk, through the galactic center), and symmetric with respect to the galactic plane.

Orbit calculations with professional programs and models have a very important consequence. The motion of an individual star is subject to physical laws of invariance or, to use a mathematical name, *integrals of motion.* Two of these conservation laws have been known for a long time: conservation of energy and of angular momentum. For a long time these two were thought to be the only two. However, certain aspects of individual orbits point at the existence of a third integral in the mechanical properties of the galactic orbits. No physical reason for this third integral has been found as yet, and we are not even sure that it is really an exact integral connected to a physical law of invariance. It could be only a pseudo integral corresponding to a physical quantity which varies so slowly that relevant changes may only be observed after a sufficiently long time.

In this chapter we will compute orbits of individual stars in a galactic model. The results will demonstrate the arguments on which the possibility of the existence of the third integral are based. This chapter is rather complicated and should therefore only be tackled by experienced readers who have successfully explored less difficult problems.

12.2 Schmidt's Model

The mass distribution in this model of the galaxy is considered as a superposition of concentric oblate spheroids, all having the same eccentricity. The mass density is constant over each individual spheroid but changes from one to another. An oblate spheroid is defined as the surface obtained by revolving an ellipse around its minor axis. The equation of an ellipse in rectangular coordinates (x, z) is

$$\frac{x^2}{a^2} + \frac{z^2}{b^2} = 1 \tag{1}$$

where a is the semi-major axis and b is the semi-minor axis. This means that the ellipse crosses the X-axis in the two tops: $(a, 0)$ and $(-a, 0)$, and the Z-axis in two other points $(0, b)$ and $(0, -b)$. The eccentricity e is defined by the relation

$$b = a\sqrt{1 - e^2}. \tag{2}$$

The ellipse reduces to a circle if $e = 0$. Otherwise, e is always between 0 and 1 for an ellipse. The larger e is, the smaller the ratio b/a, and therefore the more oblate the ellipse.

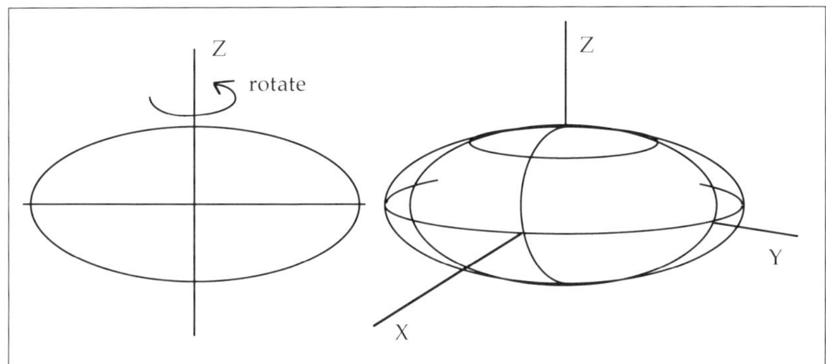

Fig. 12.1: *How to make an oblate spheroid from an ellipse.*

The oblate spheroid is obtained by revolution around the Z-axis. The surface obtained in this way has the equation

$$\frac{x^2 + y^2}{a^2} + \frac{z^2}{b^2} = 1 \tag{3}$$

or in cylindric coordinates (r,θ,z) and with (2):

$$r^2 + \frac{z^2}{1 - e^2} = a^2. \tag{4}$$

This equation is independent of the angular coordinate θ, corresponding to the symmetry around the vertical Z-axis. It is also clear that we may replace the coordinate z by $-z$ without any effect on the equation of the spheroid. This means that the oblate spheroid is symmetrical with respect to the galactic plane $z = 0$. The shape of the oblate spheroid is therefore circular for an observer placed on the Z-axis above or under the spheroid. A second consequence of the symmetry around the Z-axis is that any observer in the galactic plane outside the spheroid will observe it as an ellipse, and even observers at the same distance will see the same ellipse, independent of their galactic longitude.

In Schmidt's model, the galaxy is considered as a continuous superposition of such oblate spheroids with equal eccentricity (Fig. 12.2). The mass density is constant over each spheroid. Since every point p lies on a certain spheroid passing through it, it is always possible to find the density in p by determining the spheroid to which it belongs. In this way it is not necessary to specify the density law at every point of the model. Since all the spheroids have the same shape, the semi-major axis is the only parameter to distinguish between two spheroids. It is therefore sufficient to define the mass density law as a function of the semi-major axis.

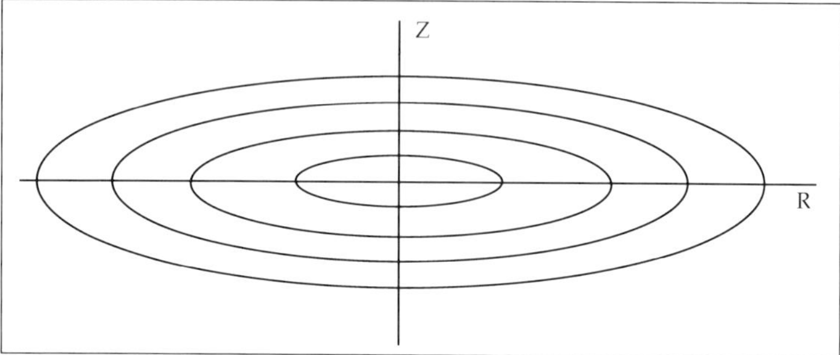

Fig. 12.2: *The superposition of concentric oblate spheroids all with the same eccentricity e.*

One would normally expect a mass density dropping to zero at the outer oblate spheroid representing the boundary of the model. Schmidt has considered a number of such distributions. In that case we would need two sets of equations of motion, one inside and one outside the galactic disk. To keep the mathematical description as simple as possible, we have decided to work with an infinite model, in which the mass density becomes negligibly small at large distances from the galactic center without actually ever becoming zero. We have, of course, selected a law yielding a finite total mass for the galaxy. This density law of our model as a function of the semi-major axis is

$$\rho(a) = \rho_C \exp\left(-\frac{a}{a_0}\right) \tag{5}$$

where ρ_C is the central density at $a = 0$ and a_0 is a scale factor.

This negative exponential law (base e) gives the density for every point on the three-dimensional oblate spheroid with semi-major axis a. The two free parameters ρ_C and a_0 are used to fit the density and some other characteristics of the model as well as possible to the observations. It is obvious that for this density law, the mass density $\rho(a)$ goes to zero when the semi-major axis a goes to infinity. This mass law was adopted instead of the law of Schmidt's model more on mathematical than on physical grounds.

The units in which we will work are the units normally used for galactic dynamics. They are as follows:

- The unit of length is 1 kiloparsec (kpc) $= 3.085 \times 10^{16}$ km.
- The unit of velocity is 1 km/s.
- The unit of time is then defined by dividing the unit of distance by the unit of velocity. This gives a time unit of 9.778×10^8years $= 3.085 \times 10^{16}$ seconds.
- The unit of mass, by requiring that the gravitational constant G equals one in galactic units, becomes 2.31×10^5solar masses $= 4.6 \times 10^{38}$ g.
- The unit of density is then the unit of mass divided by the third power of the unit of length, which yields 1.573×10^{-26}g/cm^3.

From now on we will express all our physical quantities and variables in galactic units. The definitions above may eventually be used to recalculate some of the results in normal terrestrial units.

12.3 Schmidt's Model Modified with an Infinite Density Law

The specific choice of the three free parameters e, ρ_C, and a_0 has only a quantitative effect on the final orbits. The qualitative conclusions are not affected by any reasonable change of the parameters. All results of this chapter were obtained with the following parameters:

- eccentricity of the oblate spheroid model: $e = 0.99$
- central density (in galactic units): $\rho_C = 11613.5$
- scale parameter (in kpc): $a_0 = 2.8$

Let us now evaluate the effects of these choices on some general properties of our model.

12.3.1 The Total Mass

The total mass is the sum of the masses of all the infinitesimally thin oblate spheroids. It is calculated as the three-dimensional integral of the density over the volume occupied by the model. After integration of the coordinates θ and z, one finds

$$M = 4\pi\sqrt{(1-e^2)} \int_0^\infty a^2 \rho(a) da. \tag{6}$$

We may then insert the density law specified in equation (5). The integral obtained in this way may be solved by partial integration until the term a^2 has vanished. The total mass is then

$$M = 8\pi\sqrt{1-e^2}\rho_C\, a_0^3. \tag{7}$$

The total mass with the choices for the free parameters specified above is 9×10^5 mass units or about 2×10^{11} solar masses. The density in the solar neighborhood is 326.5 density units or 5×10^{-24} g/cm^3.

12.3.2 The Rotational Curve

The gravitational field of the galactic model is completely defined by the selection of the density distribution. We are now able to continue by computing the angular velocity function at various distances from the center. This computation is performed by demanding that the gravitational acceleration towards the galactic center at every point be balanced by the centrifugal acceleration away from the center. With our specific mass density law (5) and the values for the parameters specified above we obtain the following results:

d(kpc)	v(km/s)	$P(10^6$yrs)	d(kpc)	v(km/s)	$P(10^6$yrs)
1	101	60.8	10	305	201.4
2	175	70.2	13	283	282.4
3	228	80.9	16	257	381.8
5	288	106.6	20	228	540.1
7.5	311	148			

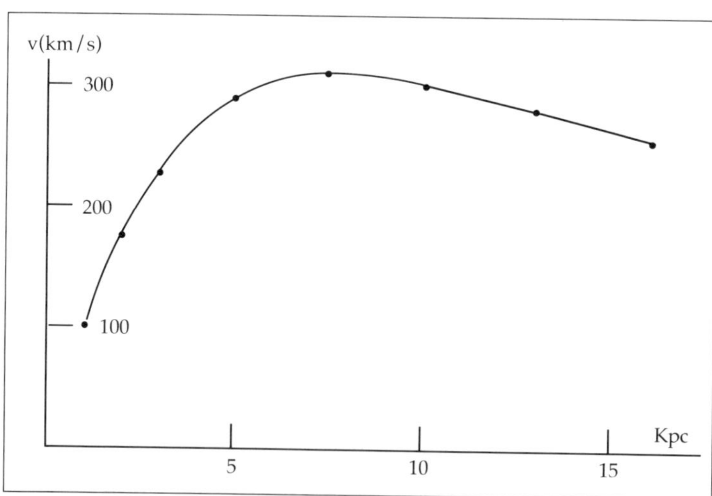

Fig. 12.3: *The rotational velocity as a function of the distance.*

12.4 The Motion of an Individual Star

12.4.1 The Equations of Motion

The gravitational force acting on a test mass representing a star has two components: a radial component directed towards the galactic center and a vertical component parallel to the Z-axis. General expressions for the corresponding accelerations in both directions have been derived in the form of integrals. When Schmidt's formulae are converted to our mass distribution (5) and our choices of the free parameters, the accelerations at a point (r, θ, z) in the galactic model are

$$K_r = -4\pi \frac{\sqrt{1-e^2}}{e^3} \rho_c r \int_0^{f(e)} \sin^2(b) \exp^{-A/a_0} db \tag{8}$$

and

$$K_z = -4\pi \frac{\sqrt{1-e^2}}{e^3} \rho_c z \int_0^{f(e)} \tan^2(b) \exp^{-A/a_0} db \tag{9}$$

where

$$f(e) = \arcsin(e) \tag{10}$$

and

$$A = \frac{1}{e} \sqrt{r^2 \sin^2 b + z^2 \tan^2 b}. \tag{11}$$

Schmidt's equations are more complicated since they must be able to distinguish between point inside and points outside the galactic disc.

All goniometric functions (sin, tan, and arcsin) should, of course, be calculated in radians. The two integrals will be solved numerically by means of Simpson's method, described in Chapter 1 (see also section 12.5).

These formulae, taking only the gravitation into account, show that the radial component is directed towards the center since its expression starts with a minus sign and all the other terms in that formula are always positive. Furthermore, the vertical component is always directed to the galactic plane. If the z coordinate is positive, the minus sign in (9) directs K_z down; if z is negative, the minus sign yields a positive (upward) acceleration. The value of $f(e)$ given by (11) turns out to be $1.4292567\ldots$ with our choice of the eccentricity $e = 0.99$. The acceleration in the Z direction is proportional to $-Z$ itself. This means that if we place a test mass in the galactic plane, with no vertical velocity component, this mass can never leave the galactic plane. Of course, this situation is not possible in reality since small details in the mass distribution of the galaxy will always affect the orbit of an individual star. The motion will in general be some kind of oscillation above and beneath the galactic plane, combined with a rotation around the rotational axis of the galactic model if the initial velocity contains both a radial, a tangential, and a vertical component.

12.4.2 Operational Framework

The orbits of the individual stars are three-dimensional, which makes them difficult to represent graphically. There are various methods available to visualize the orbits in two dimensions on a sheet of paper, a computer display or a plotter listing. An interesting way to do so is to operate in a fixed vertical plane perpendicular to the galactic plane. We may plot in that plane the points where the orbit of the star passes at every revolution around the galaxy. Such an approach demands a very large number of revolutions of the star before meaningful results can be obtained. Furthermore, a large number of iterations are needed to compute one single revolution. Since we work with rather simple numerical algorithms, neglecting the propagation of rounding errors, the accuracy is not sufficient to calculate so many iterations. Another approach, more-or-less of the same type is to plot only the r coordinate and the z coordinate as functions of the time in an RZ-coordinate system. In fact, we then follow the star in a vertical plane always containing the Z-axis and passing at every moment through the star. This vertical plane co-rotates with the star with the same angular velocity. This approach gives information concerning the distance of the star to the galactic rotational axis and the distance above or under the galactic plane. The fact that we don't have any information on the angular coordinate θ is not important since the model itself is symmetric around the vertical Z-axis and hence independent of θ.

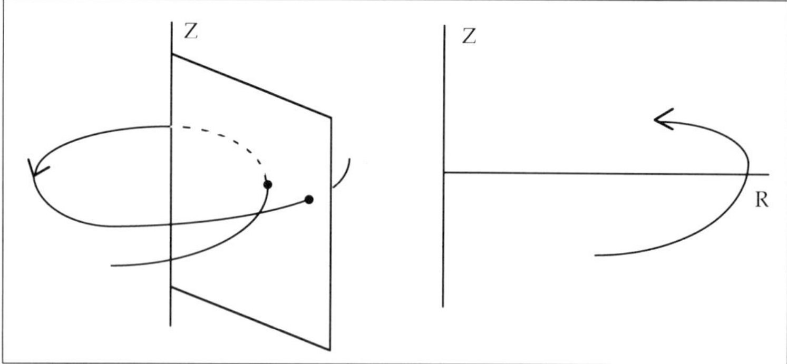

Fig. 12.4: *Left: passages of the star through a fixed plane. Right: orbit in a co-rotating vertical plane.*

We will continue in such a co-rotating plane. The effect of the angular coordinate may then be eliminated from the theory not only by plotting the results in the co-rotating vertical plane, but also by actually computing the orbit in the (r, z) coordinates. Mathematically, this means that the R-axis is no longer fixed in a certain direction, but co-rotates with the star and the vertical plane. This transformation results in the appearance of pseudo forces in the equation of motion for the R direction. This equation will not only contain the gravitational acceleration given in (8) but also a term for the centrifugal acceleration away from the galactic center. The co-rotation of the radial axis has no effect on the equation of motion in the Z direction. The new equations of motion are

$$r'' = K_r + \frac{v_t^2}{r}$$

and

$$z'' = K_z$$

where v_t is the tangential velocity component. The expression for r'' may further be transformed by applying the law of conservation of angular momentum, saying that the quantity

$$h = r\, v_t \tag{12}$$

must remain constant at every point of the orbit. The final equations of motion are then

$$r'' = K_r + \frac{h^2}{r^3} \tag{13}$$

and

$$z'' = K_z \tag{14}$$

in which the expressions for K_r and K_z are found in (8,9). Since the angular momentum h is constant, we are able to compute it once and for all at the start of the iterations when the initial conditions (position and velocity for r and z) are selected. It is then used in (13) in order to be sure of the conservation of angular momentum and to eliminate the variable v_t from the equations. Equations (13,14) are the equations we will solve to obtain the orbit of the individual star in the (r, z) coordinates of a co-rotating plane. They are only valid in such a plane because of the presence of the centrifugal acceleration term.

12.4.3 Conservation of Energy

The law of conservation of angular momentum is incorporated in the equations of motion. The second conservation law, for the total energy, may be used to check from time to time the accuracy of the calculations. The total energy of the test mass representing the star has two components: the potential energy in the gravitational field of the galaxy and the kinetic energy resulting from the motion of the star itself. Expressions for the potential energy were derived by Schmidt. When adapted to our problem with our simpler mass distribution law and infinite galactic disc, these expressions are still very complicated. We have therefore restricted them to the galactic plane, i.e., where $z = 0$. Even in that case, the expression for the potential energy $V(r, z = 0)$ is still difficult, and we will therefore not present it. It may be solved only by numerical integration methods.

Fortunately, we don't need to include this formula in our program. During our test calculations we have computed the total energy from time to time very carefully when the star was passing through the galactic plane. This was performed for a number of typical types of orbits after about 500, 1000, and 2000 iterations. The relative error on the total energy E was always less than 1%, mostly a few times 0.1%. This means that we can calculate orbits over a few thousands of iterations while maintaining reasonable accuracy, at least for our needs.

12.5 Numerical Method

The equations to be solved are the two second-order differential equations (13,14) for the motion in a co-rotating vertical plane. We first transform them into a system of four first-order equations by considering the two velocity components as explicit variables. Let us use u and v for the components of velocity in the R and Z direction. Starting from the position (r, z) and the velocity components (u, v) at time i, we compute the new position and velocity at $i+1$ after a time step Dt using our well-known midpoint method from

$$r_{i+1/2} = r_i + 0.5Dt\, u_i$$

$$z_{i+1/2} = z_i + 0.5Dt\, v_i$$

$$u_{i+1/2} = u_i + 0.5Dt\left(K_r(r_i, z_i) + \frac{h^2}{r_i^3}\right)$$

$$v_{i+1/2} = v_i + 0.5Dt\, K_z(r_i, z_i)$$

and finally

$$r_{i+1} = r_i + Dt\, u_{i+1/2}$$

$$z_{i+1} = z_i + Dt\, v_{i+1/2}$$

$$u_{i+1} = u_i + Dt\left(K_r\big(r_{i+1/2}, z_{i+1/2}\big) + \frac{h^2}{r_{i+1/2}^3}\right)$$

$$v_{i+1} = v_i + Dt\, K_z\big(r_{i+1/2}, z_{i+1/2}\big)$$

$$t_{i+1} = t_i + Dt.$$

In this way we proceed from the situation at time i to the new situation at time $i+1$, a time step Dt later. The time step Dt should be selected correctly in order to obtain small but meaningful variations of the position and the velocity components. All computations performed by the author's program and presented in section 12.6 were executed with a time step $Dt = 0.001$ in galactic units.

The equations of motion still contain the integral expressions for K_r and K_z. This means that we have to evaluate these integrals twice each time step—a first time to compute them at half the step $(i+1/2)$ and a second time for the new step itself $(i+1)$. Therefore, the numerical computation of K_r and K_z should be placed in a subprogram. This subprogram has a position (r, z) and the constant free parameters as input, and the gravitational accelerations K_r and K_z as output. An algorithm for this subprogram to compute the two integrals simultaneously is presented below. It calculates K_r and K_z over 20 steps between the boundaries of the integral 0 and $f(e)$.

Input:

- position components r, z
- free parameters e, ρ_C, a_0

Initialize:

- $K1 = 0; \;\; k2 = 0$
- $db = \dfrac{f(e)}{20}$
- $b1 = 0; \;\; b2 = db; \;\; b3 = 2db$
- $FR1 = 0; \;\; FZ1 = 0$

Do 10 times:

- $a2 = \dfrac{1}{e}\sqrt{r^2 \sin^2 b2 + z^2 \tan^2 b2}$
- $a3 = \dfrac{1}{e}\sqrt{r^2 \sin^2 b3 + z^2 \tan^2 b3}$
- $FR2 = \exp\left(\dfrac{-a2}{a_0}\right)\sin^2 b2$
- $FZ2 = \exp\left(\dfrac{-a2}{a_0}\right)\tan^2 b2$
- $FR3 = \exp\left(\dfrac{-a3}{a_0}\right)\sin^2 b3$
- $FZ3 = \exp\left(\dfrac{-a3}{a_0}\right)\tan^2 b3$

- $K1 = K1 + db\,\dfrac{FR1 + 4FR2 + FR3}{3}$
- $K2 = K2 + db\,\dfrac{FZ1 + 4FZ2 + FZ3}{3}$
- $b1 = b1 + 2db;\;\; b2 = b2 + 2db;\;\; b3 = b3 + 2db$
- $FR1 = FR3;\;\; FZ1 = FZ3$

After 10 such loops:

- $K_r = -4\pi\dfrac{\sqrt{1-e^2}}{e^3}\rho_C r K1$
- $K_z = -4\pi\dfrac{\sqrt{1-e^2}}{e^3}\rho_C z K2$

We are able to compute in this way the two integrals for any position (r, z). In each of the 10 loops, the flowchart computes the contributions of two more steps in Simpson's method for numerical integration. The fact that we compute both integrals simultaneously means that common parts of the two integrals are computed only once.

Computing the integrals needed to determine the two force components is the main reason that the calculation of an orbit takes so much computer time. About one position per second was obtained with a 10 Mhz XT computer. Therefore, the use of a mathematical coprocessor is certainly justified and will speed up the program by a factor ten or more.

12.6 Initial Conditions and Practical Hints

The initial conditions necessary to start the orbit at time $t_0 = 0$ are as follows:

- r_0 and z_0: the initial positions in the co-rotating RZ-plane (in kpc).
- u_0 and v_0: the initial velocity components in the R and Z direction (in kilometers/second).
- vt_0: the initial tangential velocity component, which is only used to obtain the angular momentum constant from

$$h = r_0\, vt_0.$$

In all the examples computed in the following section, we will select rather high velocities around a few hundred kilometers/second. The results, however, are qualitatively the same for lower speeds. Only the occupation areas will be smaller.

It may be useful to start from initial conditions in which $z_0 = 0$ and $u_0 = 0$. In that case, we start from an initial position in the galactic plane in a direction perpendicular to the position vector. In these circumstances, any position (r_i, z_i) at a positive time t_i corresponds to another point on the same orbit $(r_i, -z_i)$ which was occupied at a negative time $-t_i$, i.e., before the iteration was started. By calculating the normal future orbit from $t = 0$ with positive time steps we automatically obtain the orbit of the past for negative time steps. Although this provides no actual new information, it gives us orbit plots that are symmetric with respect to the R-axis in the co-rotating plane. In some of our practical examples, we have started from $z_0 = 0$ and $u_0 = 0$ and plotted the computed points (r_i, z_i) and the corresponding points $(r_i, -z_i)$. The successive positions (r_i, z_i) are then connected to draw the orbit, as are the negative points. Both parts of the orbit, the future and the past, meet at the initial position on the R-axis. However, this procedure must not be done unless both z_0 and u_0 are zero!

Although the plot of the orbit is two-dimensional, the real motion takes place in three dimensions. We don't compute the angular velocity component at every time step; however, it may easily be obtained using the law of conservation of angular momentum. If the angle around the Z-axis at time t_i is θ_i then the angle at t_{i+1} is found from

$$\theta_{i+1} = \theta_i + \left(\frac{Dt\ h}{r^2_{i+1/2}} \right).$$

This formula should be applied from the initial position at $t = 0$. A plot of the points (r_i, θ_i) then gives us the projection of the orbit on the galactic plane. Orbits always stay between two concentric circles around the galactic center. These circles are a result of the laws of conservation of energy and angular momentum. When z_0 and u_0 are zero, the points $(r_i, -\theta_i)$ can also be plotted (see section 7.3).

12.7 Practical Examples and Applications

All orbits in this section are computed with a time step $Dt = 0.001$ and with the parameters $e = 0.99$, $\rho_C = 11613.5$, and $a_0 = 2.8$. These parameters completely describe the structure of the galaxy model used.

12.7.1 Detailed Results for an Orbit

We first present the detailed positions computed for the orbit with the following initial conditions:

$$\begin{aligned} r_0 &= 10.00 \\ z_0 &= 0.00 \\ u_0 &= 0.00 \\ v_0 &= 150.00 \\ vt_0 &= 180.00 \end{aligned}$$

This means that $h = 1800$ and $h^2 = 3.24 \times 10^6$.

The iterations from 0 to 10 and from 101 to 110 are shown below.

i	r	z	u	v
1	9,9970	0,1500	−6,0570	149,6656
2	9,9879	0,2993	−12,0845	148,6745
3	9,9728	0,4474	−18,0551	147,0633
4	9,9518	0,5935	−23,9444	144,8856
5	9,9249	0,7371	−29,7324	142,2046
6	9,8923	0,8779	−35,4031	139,0851
7	9,8541	1,0153	−40,9443	135,5887
8	9,8104	1,1491	−46,3473	131,7698
9	9,7614	1,2789	−51,6056	127,6749
10	9,7072	1,4044	−56,7148	123,3418
.	.	.	.	.
.	.	.	.	.
.	.	.	.	.
100	9,3233	0,3324	64,2541	161,6751
101	9,3845	0,4932	58,2021	159,5318
102	9,4397	0,6515	52,2722	156,7442
103	9,4890	0,8067	46,4888	153,4359
104	9,5326	0,9584	40,8642	149,7192
105	9,5707	1,1062	35,4052	145,6891
106	9,6035	1,2498	30,1138	141,4225
107	9,6310	1,3890	24,9886	136,9790
108	9,6534	1,5238	20,0262	132,4042
109	9,6710	1,6538	15,2216	127,7317
110	9,6839	1,7792	10,5694	122,9862

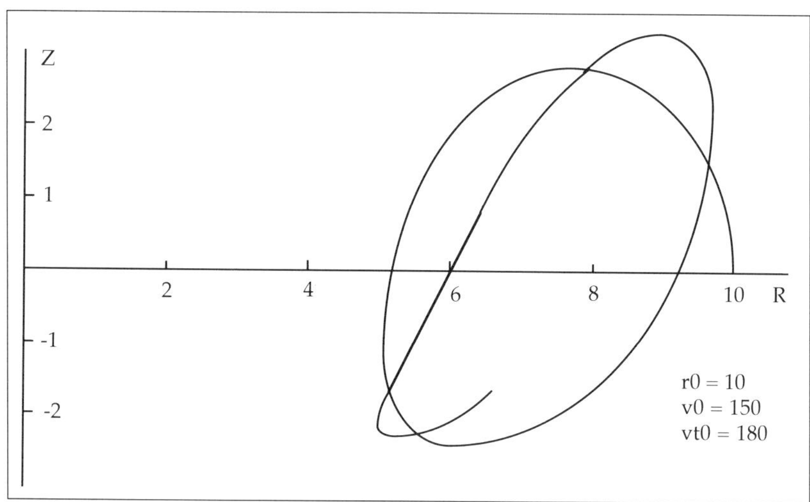

Fig. 12.5a: *Orbit in the co-rotating RZ-plane. R and Z are expressed in kiloparsec.*

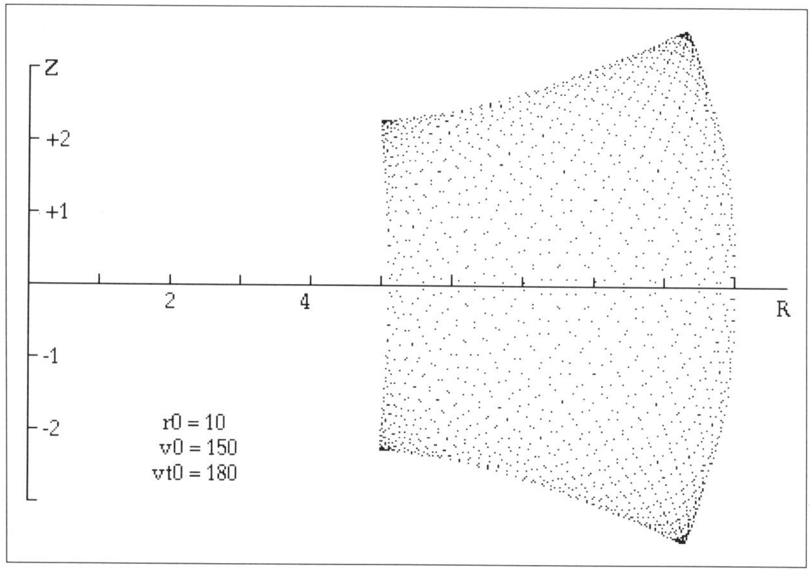

Fig. 12.5b: *Occupation area of the same orbit for 1000 iterations in the future and in the past. R and Z are in kiloparsec.*

The first 200 iterations are shown in Fig. 12.5a. Since the initial conditions for z_0 and u_0 are zero, their symmetric positions $(r_i, -z_i)$ corresponding to the past could also be plotted if desired. This is the orbit in the co-rotating

RZ-plane. Figure 12.5b shows the plots for the iterations -1000 to 1000 in the RZ plane, using (r_i, z_i) and $(r_i, -z_i)$. After a very long time period, the star will have occupied every position inside the occupation box.

The time needed to perform one revolution around the galactic center, i.e., from $\theta = 0$ to $\theta = 2\pi$, is between 165 and 166 iterations. When we use a linear interpolation, we find 165.423 iterations. Since one iteration takes 0.001 time units, a single revolution takes 0.165 time units, which is about 161.75 million years. The period of revolution around the galaxy of the Sun is about 250 million years.

12.7.2 Types of Orbits

Numerical calculations by Ollongren and Torgard (1962, *Bull. Inst. Astron. Netherlands*, **16**, 241) revealed the existence of various types of orbits in the co-rotating RZ-plane. These orbits are restricted by two laws of conservation—one for the total energy and one for the angular momentum. The latter is explicitly included in our equations of motion. Since the total energy (the sum of potential and kinetic energy) is constant, we may determine the points where the potential energy is maximal, i.e., where the kinetic energy is zero. In this way we obtain a boundary in the RZ-plane where the potential energy is so high that a negative kinetic energy would be needed to cross that boundary. Of course, this is impossible since the orbit will therefore never leave the boundary region. If no conservation law other than the two mentioned above exists, the star will have occupied all positions inside the region after a sufficiently long time. Since the results show that this is not the case, but that only a certain part (the occupation area) inside the boundary is filled, an additional restriction such as a third conservation law should exist. We are able with our program to reproduce such occupation areas of orbits. A large number of iterations is needed to compute a relevant part of the orbit. Computations of stellar orbits in the galaxy, even with our simple model, take considerable time. This chapter is therefore not intended for programmable pocket calculators.

- Box Orbits

  Our example in section 7.1 is such an orbit (Figs. 12.5a, 12.5b). In this case, the orbit will finally fill the whole box. The resulting occupation area touches the boundary of energy conservation at a number of corners, but is smaller than the total energy area. The initial conditions used for this orbit are as follows:

$$
\begin{aligned}
r_0 &= 10.00 \\
z_0 &= 0.00 \\
u_0 &= 0.00 \\
v_0 &= 150.00 \\
vt_0 &= 180.00
\end{aligned}
$$

- Ring Orbits

  The orbit fills a ring leaving a certain area in the middle empty (Figs. 12.6a, 12.6b). In the limit of this area shrinking to zero, we obtain a box orbit. The orbit for iterations 0 to 200 is plotted in Fig. 12.6a. The positions of iterations -1000 to 1000 are plotted in Fig. 12.6b. The initial conditions used for this orbit are as follows:

$$
\begin{aligned}
r_0 &= 10.00 \\
z_0 &= 0.00 \\
u_0 &= 0.00 \\
v_0 &= 180.00 \\
vt_0 &= 150.00
\end{aligned}
$$

  The occupation area still has a certain width. Try with $v_0 = 175$ to obtain a very narrow orbit.

- Tube Orbits

  The tube orbit stays in a more-or-less narrow region. When the limit of the width approaches to zero, we have a line orbit. Such a line orbit is presented in Fig. 12.7, in which the first 200 iterations are shown. The initial conditions for the examples of Fig. 12.7 are as follows:

$$
\begin{aligned}
r_0 &= 10.00 \\
z_0 &= 0.00 \\
u_0 &= 0.00 \\
v_0 &= 121.00 \\
vt_0 &= 180.00
\end{aligned}
$$

  Such orbits are very sensitive to the values of the initial conditions.

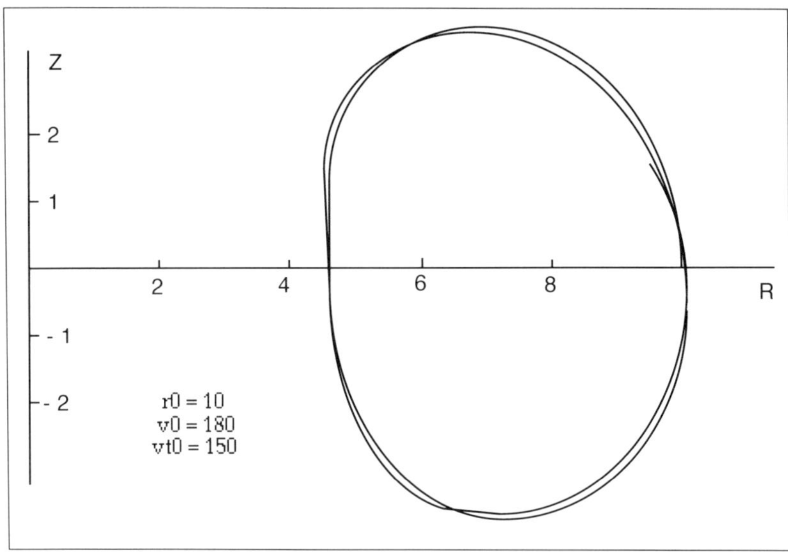

Fig. 12.6a: *Example of a ring orbit. R and Z are in kiloparsec.*

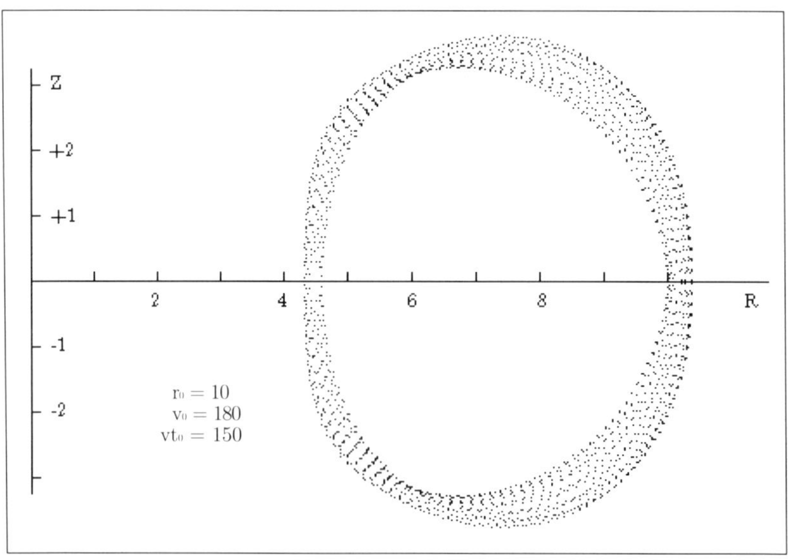

Fig. 12.6b: *Occupation area of the same ring orbit of Fig. 12.6a. R and Z are in kiloparsec.*

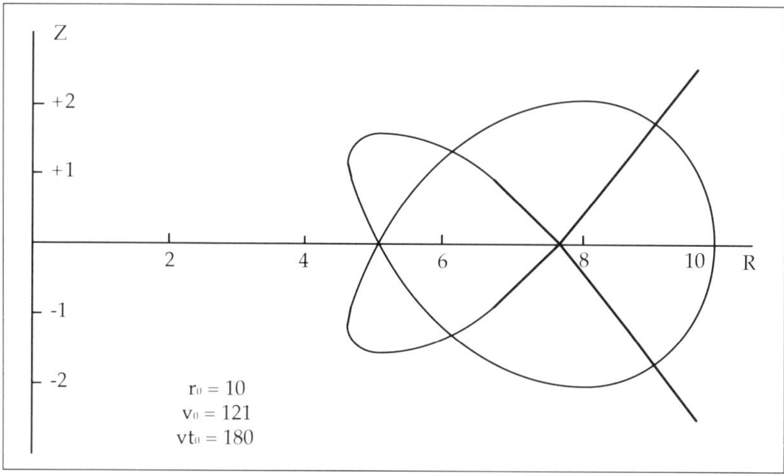

Fig. 12.7: *Example of a tube orbit. The star stays on the line and oscillates between the two boundary points (upper and lower right corner). See section 7.2 for the initial conditions. R and Z are in kiloparsec.*

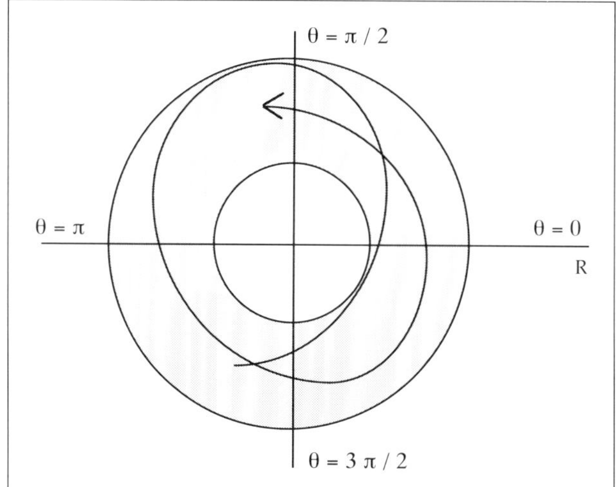

Fig. 12.8: *Occupation area projected on the galactic plane.*

12.7.3 Motion around the Rotational Axis of the Galaxy

It is also possible to plot the projection of the orbit on the galactic plane using the coordinates (r, θ). The results are much the same for all types of orbits: the star stays between two circles. These two boundaries are a consequence of the conservation laws of energy and angular momentum. Unless very specific conditions are selected, the star will have occupied every place between the two circles after a sufficient number of iterations. A typical example is presented in Fig. 12.8.

12.8 The Program Listing

The following sample program is written in MicroSoft QuickBasic 4.5

```
DECLARE SUB effes (r, z, e, fe, a0, dc, kr, kz)


CONST pi = 3.1415926536#

CLS
PRINT " Astrophysics With a PC : INDIVIDUAL STELLAR ORBITS"
PRINT " -------------------------------------------------------------"
PRINT " "
PRINT " ------------- Minimal solution program ----------------"
PRINT " "
PRINT " Input of initial conditions and parameters : "
INPUT "Initial conditions :  r(o) : ", r
INPUT "Initial conditions :  z(o) : ", z
INPUT "Initial conditions :  u(o) : ", u
INPUT "Initial conditions :  v(o) : ", v
INPUT "Initial conditions : vt(o) : ", vt0


' initialize galaxy model parameters
dt = .001
e = .99
dc = 11613.5
a0 = 2.8
fe = 1.4292567#


' initialize main cycle
h = r * vt0
past% = 0
n% = 15
PRINT "    i        t            x            z            u            v"
t = 0
ni% = 1


' here starts main cycle computing blocks of 15 iterations
DO
    FOR i% = 1 TO n%


'        results at half the step (i+/12)
          r1 = r + .5 * dt * u
          z1 = z + .5 * dt * v
          CALL effes(r, z, e, fe, a0, dc, kr, kz)
          u1 = u + .5 * dt * (kr + h * h / r / r / r)
          v1 = v + .5 * dt * kz


'        results at the full step (i+1)
          r = r + dt * u1
          z = z + dt * v1
          CALL effes(r1, z1, e, fe, a0, dc, kr1, kz1)
          u = u + dt * (kr1 + h * h / r1 / r1 / r1)
          v = v + dt * kz1
```

```
        t = t + dt

'       show newly computed iteration on screen
        PRINT USING "###### ######.#### ######.#### ######.####_
 ######.#### ######.####"; ni%; t; r; z; u; v
        ni% = ni% + 1
    NEXT i%

'   ask user whether to compute 15 iterations more or not
    PRINT "Press S to stop or any other key to continue"
    DO
        ch$ = INKEY$
        LOOP WHILE ch$ = ""
    LOOP UNTIL (ch$ = "s") OR (ch$ = "S")

END


SUB effes (r, z, e, fe, a0, dc, kr, kz)
'
'   evaluates the right hand sides of the differential equation of motion
'   for the coordinates r and z. Both right hand sides are computed in the
'   same cycle applying Simpson's rule for the numerical calculation of a
'   defined integral
'
'
     k1 = 0
     k2 = 0
     db = fe / 20
     b1 = 0
     b2 = db
     b3 = 2 * db
     fr1 = 0
     fz1 = 0

'    next FOR cycle computes ten pairs of subintervals for the numerical
'    integration with Simpson's rule
     FOR ii% = 1 TO 10
          a2 = 1 / e * SQR((r * SIN(b2)) ^ 2 + (z * TAN(b2)) ^ 2)
          a3 = 1 / e * SQR((r * SIN(b3)) ^ 2 + (z * TAN(b3)) ^ 2)
          fr2 = EXP(-a2 / a0) * (SIN(b2)) ^ 2
          fz2 = EXP(-a2 / a0) * (TAN(b2)) ^ 2
          fr3 = EXP(-a3 / a0) * (SIN(b3)) ^ 2
          fz3 = EXP(-a3 / a0) * (TAN(b3)) ^ 2
          k1 = k1 + db * (fr1 + 4 * fr2 + fr3) / 3
          k2 = k2 + db * (fz1 + 4 * fz2 + fz3) / 3
          b1 = b1 + 2 * db
          b2 = b2 + 2 * db
          b3 = b3 + 2 * db
          fr1 = fr3
          fz1 = fz3
     NEXT ii%

' multiply the two integrals with the other factors of the equations
' of motion
     kr = -4 * pi * SQR(1 - e * e) / e ^ 3 * dc * r * k1
     kz = -4 * pi * SQR(1 - e * e) / e ^ 3 * dc * z * k2

END SUB
```

Chapter 13

Cosmological Models for the Universe

13.1 Introduction

One of the most spectacular astrophysical discoveries occurred during the first half of this century—the global expansion of the universe. Between 1920 and 1930 the American astronomer Edwin Hubble studied spectra of other galaxies and found that they all show spectral lines shifted to the red, meaning that these galaxies all drift away from us. The phenomenon was observed in all directions in exactly the same way. This result became even more spectacular when it turned out that the run-away velocity of a galaxy is proportional to its distance. The larger the distance, the larger the red shift and the higher the velocity. The relation between the run-away velocity v and the distance d is

$$v = H_0 \, d. \tag{1}$$

The velocity is expressed in kilometers/second, and the distance in megaparsec (Mpc). The proportional constant H_0 was later called the Hubble constant. Its exact value has always been a problem because of the many difficulties and uncertainties in the measurements of extragalactic distances. A value given by Allan Sandage in 1972 is

$$H_0 = 55 \pm 7 \frac{\text{km/s}}{\text{Mpc}}. \tag{2}$$

For instance, a galaxy at a distance of 8 Mpc has a run-away velocity of 440 km/s. We will always adopt this value for the Hubble constant in our calculations. The results only differ *quantitatively*, not qualitatively, when another value is used. Furthermore, since the problem will be solved in dimensionless coordinates, the value of H_0 has no effect on the solution of the equations.

So we know that at the moment our universe expands. Has it always been so? Is the Hubble constant, constant in space, also constant in time? If not, was it larger or smaller in the past? And what about the future? Will the universe keep expanding forever, or will it reach a maximal size and then turn into a contraction stage? If so, how will it end?

These questions are the subject of this chapter. We will analyze the possible past, present, and future stages of our universe. The history of the universe is studied by computing the evolution of its scale factor. The behaviour of the scale factor, how it increases or decreases in time, describes the way our universe expands or contracts. Various models have been examined by well-known astronomers such as De Sitter, Friedman, Einstein, Lemaître and Milne. Unfortunately, we are, at this point in time, unable to determine in which model we are living.

13.2 Physical and Mathematical Backgrounds

In all the following formulae the subscript o stands for physical quantities at their present value, i.e., at $t = 0$. For instance, P_0 is the actual pressure in the universe. The two general formulae describing the structure and evolution of a spherically symmetric universe are

$$8\pi\rho G = \frac{3}{R^2}\left(kc^2 + \left(R'\right)^2\right) - \Lambda \tag{3}$$

and

$$8\pi\rho GPc^{-2} = -2\frac{R''}{R} - \frac{1}{R^2}\left(kc^2 + \left(R'\right)^2\right) + \Lambda. \tag{4}$$

These two fundamental equations from Einstein's theory of general relativity contain three unknown functions of time, namely, the scale factor $R(t)$, the average density $\rho(t)$, and the average pressure $P(t)$. The problem is therefore undefined because there is one variable too much or one equation missing to solve the problem.

A first possibility is to assume a polytrope-like relation linking the density and the pressure, but this leads to a problem with four free parameters, even after transformation to dimensionless coordinates. Therefore, most of the research was directed towards another possibility: the zero-pressure models. It is assumed in this solution that the pressure is zero, a condition justified by the observations. The problem then becomes easier since it is defined: we now have two variables and two equations with a total order of two. The conditions at which the total amount of mass has to remain constant during the evolution of the universe is

$$\rho_0 R_0^3 = \rho(t)R(t)^3. \tag{5}$$

With this relation and using the assumption $P = 0$, the general equations (3,4) may easily be combined to one second-order differential equation:

$$R'' = -4\pi G\rho_0 \frac{R_0^3}{3R^2} + \frac{\Lambda R}{3} \tag{6}$$

where

R is the scale factor, with its first and second derivatives R' and R'' and its present value R_0,

ρ_0 is the present average density of the universe,

G is the gravitational constant, and

Λ is the cosmological constant.

The first part of equation (6) represents the gravitational force working to keep the universe together. As usual, its intensity decreases with the square of the distance. The second part of the equation is a cosmological force. If Λ is zero, gravitation is the only force. If not, the cosmological force is directed outwards if Λ is positive and inwards if negative. The cosmological force was introduced by Einstein in order to obtain static models, although there was no physical reason to reject other possibilities. It took some years for him to accept the idea of a Big Bang, although the Big Bang was a direct consequence of his own theory of general relativity. Einstein later called the idea of a cosmological constant the biggest mistake he'd ever made.

Another way to combine (3,4) gives the equation

$$\left(R'\right)^2 = 8\pi G\rho_0 \frac{R_0^3}{3R} + \frac{\Lambda R^2}{3} - kc^2 \tag{7}$$

in which k, the curvature of the universe, appears. It can take values of -1, 0, or 1 for negative, zero, or positive curvature. A physical interpretation of this important concept will be presented in section 13.5. The equation above is often used in theoretical textbooks. We will work with (6), which may be solved more easily.

Apart from the differential equation (6), there are some additional relations between various physical parameters. Therefore, two free parameters are introduced to determine the physical characteristics of a particular model universe: the density parameter (σ_0) and the acceleration parameter (q_0).

1. The density parameter σ_0:

$$\sigma_0 = \frac{4}{3}\pi\rho_0 G H_0^{-2}. \tag{8}$$

With Sandage's value for H_0, this means that

$$\rho_0 = 1.1 \times 10^{-29} \sigma_0 \,\text{gram/cm}.$$

Of course, the density parameter, which is proportional to the average density, is strictly positive and takes values around 1.

2. The acceleration parameter q_0:

$$q_0 = -\frac{R_0''}{R_0} H_0^{-2}. \tag{9}$$

This gives the relative acceleration of the scale factor. The presence of Hubble's constant H_0 is needed to make the acceleration parameter dimensionless. Typical values are in the range -2 to 2.

These two parameters are linked with the classical parameters by the following relations (which you may try to prove as an exercise):

$$\Lambda = 3H_0^2(\sigma_0 - q_0) \tag{10}$$

$$R_0^2 = kc^2(3\sigma_0 - q_0 - 1)^{-1} H_0^{-2} \tag{11}$$

$$k = \text{Sign}(3\sigma_0 - q_0 - 1) \tag{12}$$

The curvature k is given by the sign of $3\sigma_0 - q_0 - 1$. If this quantity is zero, positive, or negative, k will be 0, +1, or -1.

The next step usually applied is to transform (6) into a form that contains only σ_0 and q_0 as free parameters. This may be performed quite easily. It will simplify not only the initial conditions of the differential equation, but also the interpretation of the results. First, we take the actual value of the scale factor as unit of length. The new variable Y is then defined by

$$Y = \frac{R}{R_0} \tag{13}$$

with $Y_0 = 1$.

The new unit of time is taken H_0^{-1} or, in other words, 1.801×10^{10} years, when the value given in (2) is used. The new time coordinate X is therefore related to the time t by:

$$X = H_0 \, t \tag{14}$$

with $X_0 = 0$.

Furthermore, with (13,14) it is easy to show that

$$Y_0' = 1. \tag{15}$$

These transformations mean that the actual value of the constant of Hubble becomes 1 in the new coordinates (X,Y). The choice of $X_0 = 0$ is made possible by the fact that the time does now appear explicitly in the equations. For instance, $X = 2.46$ means 4.43×10^{10} years in the future, while $X = -0.46$ corresponds to an event 8.28×10^9 years ago.

When (6) is transformed in the coordinates (X,Y) instead of (t,R), and with the two new free parameters σ_0 and q_0 instead of ρ_0, Λ, k and H_0 one finds that

$$Y'' = -\frac{\sigma_0}{Y^2} + (\sigma_0 - q_0)Y \tag{16}$$

which is quite a simple equation. Another advantage is the fact that the Hubble constant has disappeared from the equation. We are therefore independent of its exact value. This equation (16) is the one we will have to solve in order to obtain the evolution of the scale factor of the universe. Try to perform the transformation from (6) to (16) as an exercise. The computation of a model universe takes four steps.

1. Select values for σ_0 and q_0
2. Solve (16) numerically using a positive time step DX for the future and starting from the initial conditions $Y_0 = 1$ and $Y_0' = 1$ at $X = 0$. Eventually, the calculations will be stopped by a collapse in the far future.
3. Solve the same equation with a negative time step for the past. Eventually, we will find that the universe originated from a Big Bang.
4. Determine the physical parameters with (10,11,12).

Many cases may be solved analytically without numerical integrations. We refer to Mc Vittie's textbook for these solutions (see Bibliograghy). We will write a general numerical program able to handle any possible zero-pressure model.

13.3 Types and Classes of Zero-Pressure Models

The models of the zero-pressure collection are divided into three types and seven classes.

Type A1. General property: $\sigma_0 = q_0$ and hence $\Lambda = 0$.

Class	Parameters	k	Λ	Name
A1 i	$\sigma_0 = q_0 = 0$	-1	0	Milne
A1 ii	$\sigma_0 = q_0 < 0.5$	-1	0	
A1 iii	$\sigma_0 = q_0 = 0.5$	0	0	Einstein
A1 iv	$\sigma_0 = q_0 > 0.5$	1	0	

Type A2. General property: $\sigma_0 = 0$ (zero-density models).

Class	Parameters	k	Λ	Name
A2 v	$\sigma_0 = 0; q_0 > 0$	-1	neg	
A2 vi-a	$\sigma_0 = 0; q_0 < -1$	1	pos	
A2 vi-b	$\sigma_0 = 0; q_0 = -1$	0	pos	De Sitter
A2 vi-c	$\sigma_0 = 0; -1 < q_0 < 0$	-1	pos	

Type A3. All other zero-pressure models with $\sigma_0 > 0$ and $\sigma_0 <> q_0$. Only a few models of this type have been classified, namely the classes A3 vii-a and A3 vii-b. Models of these classes have values for σ_0 and q_0 that are certain rational functions of a third parameter n. We will not pay attention to these classes since they are less important.

Class	Parameters	k	Λ	Name
A3 vii-a		-1	neg	
A3 vii-b		1	pos	

Other models such as the hesitating model of Lemaître, named after the father of the Big Bang idea, are not divided into classes but belong to type A3. The models in which pressure is included need a more complicated equation than (16) and are less intensively studied because of the larger number of free parameters. Only a few of them are classified as members of type B.

Even the zero-pressure models, having only two free parameters, allow for a large number of models. One should always keep in mind, however, that the only "real" models are the ones with a cosmological constant Λ equal to zero, hence models of class A1. All these models start from a Big Bang since gravitation is the only large scale force acting in such a universe. Since there is only one universe, or at least one universe that is visible to us, there should be only one good model. The fact that the theory still allows fundamentally different A1-types of solutions indicates that the theory is not good enough to predict the evolution of the universe correctly.

13.4 Numerical Method

We have to solve (16) in a numerical way. The calculation itself has two parts. First, starting from $X_0 = 0$ with the initial conditions $Y_0 = 1$ and $Y_0' = 1$, the future is computed with positive time steps. Afterwards, starting from the same initial conditions, the past is calculated with negative time steps. Since we compute both the past and the future from the same initial conditions, a smooth transition at $X = 0$ is guaranteed. Let us now call the first derivative of Y simply Z, instead of Y'. In this way Z may be considered

as the expansion velocity of the model. The second-order equation is then transformed into a system of two coupled first-order equations which may be solved easily. Starting from a point (X_i,Y_i) the next point $i+1$ is computed as follows:
First

$$X_{i+1/2} = X_i + 0.5DX \tag{19}$$

$$Y_{i+1/2} = Y_i + 0.5DX\, Z_i \tag{20}$$

$$Z_{i+1/2} = Z_i + 0.5DX\left(-\frac{\sigma_0}{Y_i^2} + (\sigma_0 - q_0)Y_i\right) \tag{21}$$

then

$$X_{i+1} = X_i + DX \tag{22}$$

$$Y_{i+1} = Y_i + DX\, Z_{i+1/2} \tag{23}$$

$$Z_{i+1} = Z_i + DX\left(-\frac{\sigma_0}{Y_{i+1/2}^2} + (\sigma_0 - q_0)Y_{i+1/2}\right) \tag{24}$$

starting from $X_0 = 0$, $Y_0 = 1$, and $Z_0 = 1$.

13.5 Some Remarks

13.5.1 On the Choice of DX

A time step between 0.02 and 0.05 may be used during the largest part of the history of the model. However there are two situations in which DX should be decreased to a smaller value. These two events are a Big Bang and a collapse. For instance, all the models of type A1 and some of A2 and A3 originate from a Big Bang, and some models (such as A1 iv and A2 v) evolve into a collapse in their far future. In both cases Y becomes zero, which means that in (16) Y'' becomes infinite. Furthermore, the first derivative Y' becomes too large for our program to handle. Therefore, the time step should be decreased when one of the two events mentioned is reached. A possible practical method is to reduce the time step to 0.01 as soon as Z is larger than 2 or more negative than -2.

One should always check $Y_{i+1/2}$ and Y_i on being negative. As soon as one of them is negative, any further calculation is meaningless.

13.5.2 An Interpretation of the Curvature k

The value of k (-1, 0, or 1) is not known for our universe; however, it should be possible, at least in theory, to derive k from observations of the number density of extragalactic systems as a function of their distance to the Earth. This idea is easy to explain in two dimensions.

Curvature of space is a strange thing. You cannot really see it since the bearers of light, the photons, follow the curvature; thus, space looks straight. Suppose we have three two-dimensional universes, one with a negative curvature (right), one with zero curvature (middle), and one with positive curvature (left). All three of them are homogeneously filled with galaxies so that there is no place in the universe where the average number of galaxies per square unit of length is larger or smaller than in the rest of the universe.

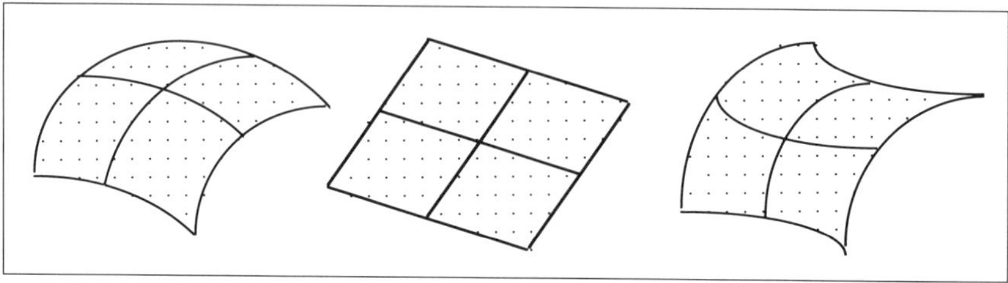

Since the photons will follow the curvature of space, each observer will see his space as flat. For instance, if an observer emits a beam of light, he will see it as a straight line. To reconstruct how observers see the number density of galaxies in their respective spaces, we have to make the three universes flat. When this is done for the case of positive curvature ($k = 1$), the space will split at the borders, just as if half an orange peel is squeezed flat on a table. The universe with $k = 0$ remains unchanged since it is already flat, and for $k = -1$, we will have *too much space* at the borders, making it necessary to fold it a little bit.

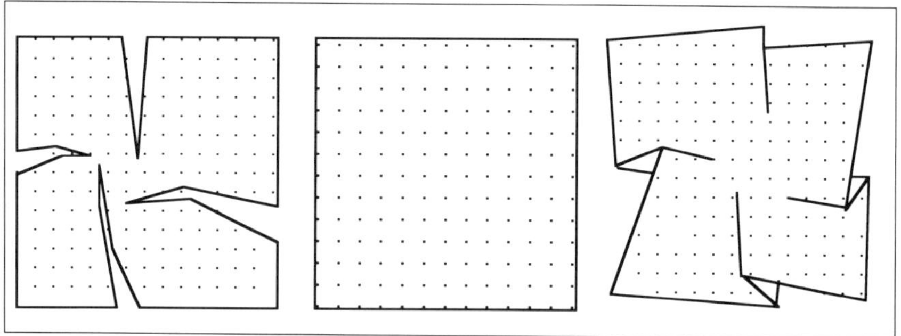

To make the splitting in $k = 1$ disappear, the space there should be stretched out so that the density of galaxies at the borders (i.e., at large distance) will decrease. For $k = 0$, the density remains constant and unchanged.

For $k = -1$, the space at the borders should be compressed to make the foldings disappear, so that the number density of galaxies will increase there.

The observer in the positively curved space will see that the number density of galaxies decreases as a function of the distance, while for the observer in $k = -1$ the number density will increase. The same phenomenon is present in a three-dimensional universe, but it is not possible to draw these spaces on a sheet of paper. Therefore, a plot of the number density of galaxies in our universe as a function of the distance should inform us on the curvature of our universe. In reality, it is not possible to use galaxies since they are grouped in clusters and superclusters.

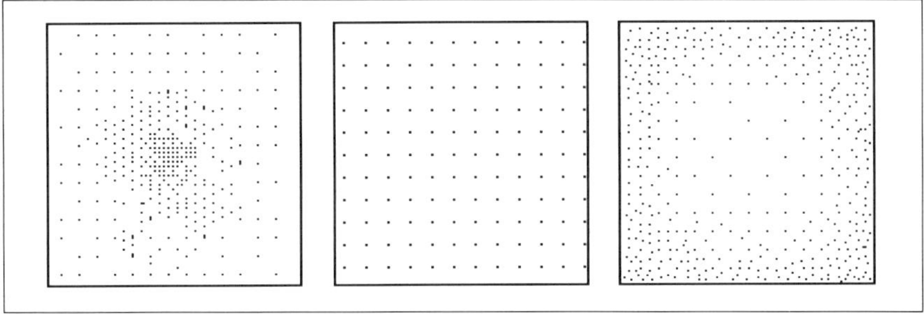

Unfortunately, this method is only theoretical since the uncertainties of the observations and the very large distance needed to observe a variation of the number density do not allow a successful application.

13.6 Practical Examples

We will present in this section a brief description of the models of types A1 and A2, illustrated with computations performed to test your program. The time step used is always 0.02 or -0.02.

13.6.1 Type A1

Since all these models have the two free parameters σ_0 and q_0 equal to each other, the cosmological constant and therefore the cosmological force are zero. These models are only governed by gravitation. The expansion rate of such models can only decrease, since gravitation tends to contract the model.

A1 i This is the most simple case since both basic parameters are zero. This means that there are no forces active in this model, which will keep expanding forever at the same rate. The solution is a straight line in the XY-plane with the equation $Y = X$.

A1 ii In this model σ_0 is small, thus corresponding to a small gravitational effect. The expansion velocity will slow down but not enough to turn the expansion into a contraction. Therefore, when the model becomes infinitely large, the expansion velocity is still positive. As a test for your program we present some parts of the iterations or a model with both σ_0 and q_0 equal to 0.35. The time step is always 0.02, even when $Z = Y'$ varies steeply.

Future

$X =$	$Y =$	$Y' =$
0.00	1.0000000	1.0000000
0.02	1.0199300	0.9931379
0.04	1.0397255	0.9865380
0.06	1.0593915	0.9801838
0.08	1.0789328	0.9740605
0.10	1.0983539	0.9681544
⋮	⋮	⋮
0.50	1.4666426	0.8816433
0.52	1.4842429	0.8785279
0.54	1.5017797	0.8752876
0.56	1.5192544	0.8722198
0.58	1.5366685	0.8692215

Past

$X =$	$Y =$	$Y' =$
0.00	1.0000000	1.0000000
−0.02	0.9799300	1.0072421
−0.04	0.9597143	1.0145840
−0.06	0.9393466	1.0223473
−0.08	0.9188203	1.0304560
−0.10	0.8981283	1.0389367
⋮	⋮	⋮
−0.66	0.1672057	2.1142463
−0.68	0.1224170	2.4423569
−0.70	0.0688989	3.1713141

The reduced scale factor Y becomes negative for $X = -0.72$. We therefore conclude that this model was born from a Big Bang at about $X = -0.71$, which means 1.28×10^{10} years ago. It keeps expanding forever, although the expansion rate decreases.

A1 iii This model is the interface between A1 ii and A1 iv. The expansion rate drops to zero at the moment Y becomes infinite.

A1 iv In this class σ_0, and therefore the density, is large enough to generate a sufficient gravitational force to turn the expansion into a contraction stage before the universe is infinitely large. The model ends in a collapse. With $\sigma_0 = q_0 = 1$, for instance, one obtains the following data:

Future		
$X = 0.00$	$Y = 1.0000000$	$Y' =$ 1.0000000
0.02	1.0798000	0.9803941
0.04	1.0392156	0.9615277
0.06	1.0582609	0.9433466
0.08	1.0769493	0.9258023
0.10	1.0952929	0.9088509
.	.	.
.	.	.
.	.	.
5.68	0.1694588	−3.2716456
5.70	0.0970612	−4.3412521

At the next point $X = 5.72$ and the scale factor Y is negative. The model collapses at $X = 5.71$, which is over 1.03×10^{11} years. A maximum is reached at $X = 2.58$ when Y is twice its actual value. From then on the model starts contracting. The Big Bang occurred at $X = -0.57$.

When σ_0 is decreasing from large values to 0.5, the time of the collapse increases. For $\sigma_0 = 0.5$ the time of collapse is infinite. Therefore, A1 iii is the limit of A1 iv for σ_0 decreasing to 0.5. In the same way, it is the limit of class A1 ii for σ_0 increasing from zero to 0.5.

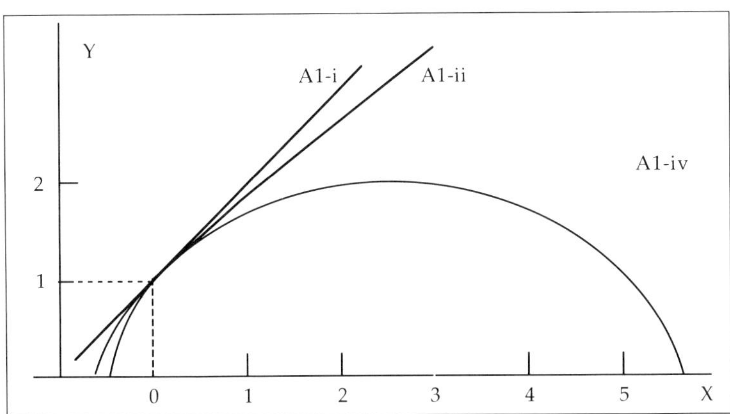

Fig. 13.1: *Some examples of Type A1 models.*

13.6.2 Type A2

In type A1 there was only gravitation; now there is only the cosmological force since $\sigma_0 = 0$. There is no matter in this universe to generate a gravitational field.

A2 v In this model q_0 is positive so that the cosmological constant itself is negative. The cosmological force aims to contract the presently expanding model and will succeed since there is no outward force in this model. Furthermore, the intensity of the contracting force increases with the distance since the cosmological factor in (6) is proportional to the scale factor R. The resulting model is very similar to the models of class A1 iv. The model reaches a maximum size and ends in a collapse.

A2 vi This class is divided into three subclasses: a, b, and c. For A2 vi-a, the solution in the XY-plane is U-shaped. The model was very large in the past, and its size decreased to a certain minimum, the bottom point of the U shape. This point lies above the X-axis. From then on the model started to grow again because of a positive value of L. When q_0 grows from negative values smaller than -1 to -1 itself, the bottom point shifts nearer to the X-axis and more and more to the left. It reaches the X-axis when q_0 equals -1. At that moment $X(\text{bottom}) = -\infty$. This is the limiting model A2 vi-b. For A2 vi-c, with q_0 between -1 and 0, the solution starts with a Big Bang and expands forever with increasing expansion velocity ($R'' > 0$).

The models of class A2-vi may easily be solved analytically, since the differential equation reduces to

$$Y'' = -q_0 Y.$$

Let us now replace q_0 by $-a$. Since q_0 is negative, a is a positive quantity. This gives'

$$Y'' = aY.$$

Verify that any function

$$Y = C1 \exp(aX) + C2 \exp(-aX)$$

with

$$C1 = \frac{a+1}{2a}$$

and

$$C2 = \frac{a-1}{2a}$$

satisfies the differential equation. For instance, for class A2 vi-b we obtain $a = 1$, which yields an exact solution:

$$Y = \exp(X).$$

The scale factor of this universe behaves like a perfect exponential function. Its age is endless, and its expansion rate keeps increasing more and more.

13.6.3 Type A3

Only a few models have been classified in the classes vii-a and vii-b. In fact, type A3 contains all the zero-pressure models in which both the gravitational and the cosmological forces are active. To illustrate the large variety of possible models the reader is invited to compute Lemaître's so-called "hesitating" model.

After a Big Bang the expansion velocity decreases during a certain time as if the model is going to reach a maximal size and then start to contract. However, before the expansion rate has dropped to zero, the cosmological force (acting as a repulsion in this case) is larger than the gravitational component. As a result, the expansion rate slowly increases again. The point where both forces are equal is where $Y'' = 0$. With $\sigma_0 = 1.5$ and $q_0 = 1.3125$, this happens when $Y = 2$. The expansion rate at that moment is $Y' = 0.25$.

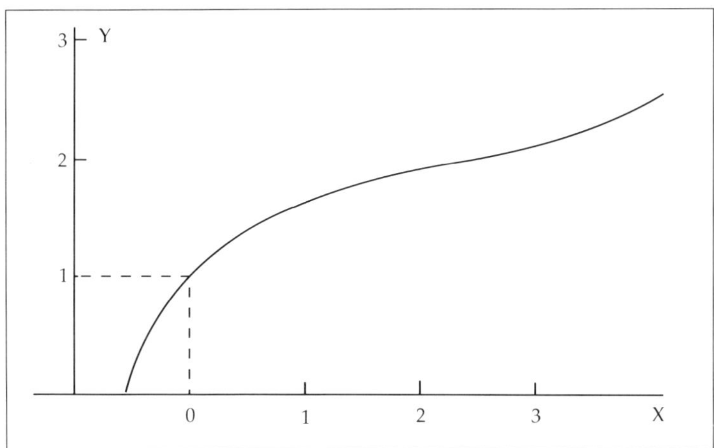

Fig. 13.2: *Lemaître's hesitating model.*

13.7 The Program Listing

The following sample program is written in MicroSoft QuickBasic 4.5:

```
CLS
PRINT "Astrophysics With a PC : UNIVERSE MODEL"
PRINT "-----------------------------------------"
PRINT " "
```

```
PRINT "--------- Minimal solution program ----------"
PRINT " "
PRINT "Input of the parameters : "
INPUT "sigma(o) : ", s
INPUT "q(o)     : ", q
PRINT " "

' this FOR computes the past (era%=0) and the future (era%=1)
FOR era% = 0 TO 1
    x = 0
    y = 1
    z = 1
    ctn% = 1

' select the time step
    IF era% = 0
    THEN
        dx = -.02
        PRINT "Computations for the past"
    ELSE
        dx = .02
        PRINT "Computations for the future"
    END IF

'  disply heading of table of results
    PRINT " "
    PRINT "    x           y          z"
    count% = 0
    PRINT USING "####.####  ####.####  ####.####"; x; y; z

'  main cycle of the iterative procedure
    DO

'      decrease time step if function y is too steep
        IF ABS(z) > 2 THEN dx = .01 * dx / ABS(dx)

'      results at half the step
        x12 = x + .5 * dx
        y12 = y + .5 * dx * z

'      check if scale factor y12 is still positive
        IF y12 > 0
        THEN
            z12 = z + .5 * dx * (-s / y / y + (s - q) * y)

'          results for the full step
            x = x + dx
            y = y + dx * z12

'          check if scale factor y1 is still positive
            IF y > 0
            THEN
                z = z + dx * (-s / y12 / y12 + (s - q) * y12)
                count% = count% + 1

'              show results of newly computed iteration on screen
                PRINT USING "####.####  ####.####  ####.####"; x; y; z
```

```
'               offer the user the possibility to stop the actual era%
'               every 15 iterations
                IF (count% + 1) MOD 15 = 0
                THEN
                    PRINT "Continue (y/n) ?"
                    DO
                    answer$ = LCASE$(INKEY$)
                    LOOP UNTIL (answer$ = "y") OR (answer$ = "n")
                    SELECT CASE answer$
                    CASE "y"
                        ctn% = 1
                    CASE "n"
                        ctn% = 0
                    END SELECT
                END IF

            ELSE                'This ELSE is reached when Y1 was negative'
                IF era% = 0
                THEN
                    PRINT "Model starts from Big Bang"
                ELSE
                    PRINT "Model ends in a Collapse"
                END IF

                PRINT "Press any key to continue"
                DO
                LOOP WHILE INKEY$ = ""
                ctn% = 0
            END IF

        ELSE                    'This ELSE is reached when Y was negative
            IF era% = 0
            THEN
                PRINT "Model starts from Big Bang"
            ELSE
                PRINT "Model ends in a Collapse"
            END IF

            PRINT "Press any key to continue"
            DO
            LOOP WHILE INKEY$ = ""
            ctn% = 0
        END IF

    LOOP UNTIL ctn% = 0

NEXT era%

PRINT "Press any key to stop the program"
DO
LOOP WHILE INKEY$ = ""
END
```

Appendix

Some Physical and Astrophysical Equations

Astronomical Unit	A.U.	=	1.49598×10^{13} cm
Gas constant	R	=	8.314×10^{7} erg/mol/K
Gravitational constant	G	=	6.673×10^{-8} cm^3/g/s^2
Hubble constant	H_o	=	55km/s/Mpc
Light-year	l.y.	=	9.4673×10^{17} cm
Luminosity of the Sun	L_o	=	3.83×10^{33} erg/s
Mass of the Sun	M_o	=	2×10^{33} g
Mean density of the Sun	ρ_o	=	1.42 g/cm^3
Parsec	pc	=	3.0859×10^{18} cm
Planck's constant	h	=	6.6252×10^{-27} erg s
Stephan Boltzmann constant	a	=	7.56464×10^{-15} erg/cm^3/K^4
Radius of the Sun	R_o	=	6.96×10^{10} cm
Speed of light	c	=	2.9979×10^{10} cm/s
Radiation density constant	σ	=	5.6687×10^{-5} erg/cm^2/K^4/s

Bibliography

General Astrophysics

Allen, C. *Astrophysical Quantities.* London: Athlone Press, 1973.

Hynec, J., ed. *Astrophysics, A Topical Symposium.* New York: McGraw-Hill, 1951.

Lang, K. *Astrophysical Formulae.* Berlin: Springer Verlag, 1974.

Rosseland, S. *Theoretical Astrophysics.* Oxford: Clarendon Press, 1973.

The Solar System

Alfvén, H., and G. Arrhenius. *Structure and Evolutionary History of the Solar System.* Dordrecht: D. Reidel Publ. Co., 1975.

Brandt, J., and P. Hodge. *Solar System Astrophysics.* New York: McGraw-Hill, 1964.

Wurm, K. *The Moon, Meteorites and Comets.* Chicago: University Press, 1963.

Celestial Mechanics

Szebehely, V. *Theory of Orbits.* New York: Academic Press, 1967.

Stellar Structure and Atmospheres

Chandrasekhar, S. *Stellar Structure.* New York: Dover Publ., 1938.

Clayton, D. *Principles of Nucleosynthesis.* New York: McGraw-Hill, 1968.

Mihalis, D. *Stellar Atmospheres.* San Francisco: Freeman, 1978.

Schwarzschild, M. *Structure and Evolution of Stars.* New York: Dover Publ., 1958.

Galaxies and Cosmology

Mc Vittie, G. *General Relativity and Cosmology.* London: Chapman and Hall, 1956.

Misner, C., K. Thorne, A. Wheeler. *Gravitation.* San Francisco: Freeman, 1973.

Index

E

F

G

H

I

J

K

L

M

N

O

P

R

S

T

Optional Software for the IBM-PC

Three versions of 12 programs in either QuickBasic or Pascal languages are presented in this software offering. The first and most simple version is source code as presented in this book (suitably modified in the case of Pascal). Next, source code that makes the programs more user friendly and in the case of 5 of the programs, they include graphic options. Finally, compiled versions of these programs are included along with a set of ASCII files showing suitable inputs so that the first time user may become familiar with the program by running it with valid inputs. The programs are as follows:

1. `ATMOS` computes a simplified stellar atmosphere.
2. `COMET` computes the shape of a comet tail in the neighborhood of the sun.
3. `COSMOS` computes a cosmological model of the universe.
4. `DWARF` computes the structure of a white dwarf.
5. `EQUIPOT` computes equipotential surfaces of the two body problem.
6. `GALAXY` computes an individual stellar orbit in the galaxy.
7. `METEOR` computes the evolution of a meteoroid entering the terrestrial atmosphere.
8. `PARALLAX` computes dynamical parallax for wide binaries.
9. `POLYTROP` computes the structure of a polytrope.
10. `STARFORM` computes the evolution of a stellar birth region in the galaxy.
11. `STARMOD` computes a composite polytrope model for a Zero Age Main Sequence star.
12. `THREEBOD` computes restricted three body problem.

Five of these programs offer, in addition to numerical results, the option of a graphic presentation of the results. These programs will run on a PC with black and white CGA (640 by 200), 4 color EGA (640 by 350) and 4 color VGA (640 by 480). The following programs have this option:

1. COMET The tail is drawn for a number of orbit positions before, during and after the perihelion passage.
2. THREEBOD The orbit of the third mass is drawn with respect to the positions of the two primary masses.
3. STARFORM Presents phase diagrams with stellar and atomic fraction versus molecular fraction as well as these three fractions as functions of time.
4. GALAXY The orbit of an individual star is drawn in a co-rotating vertical plane.
5. EQUIPOT Various options are available.

The following illustrations were produced by the above 5 programs.

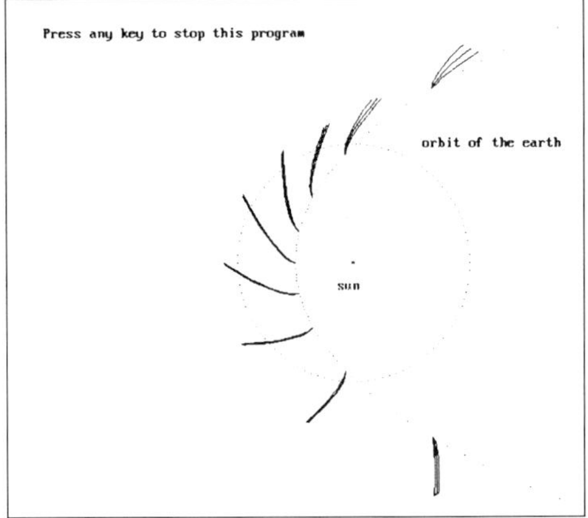

Graphic presentation from COMET.

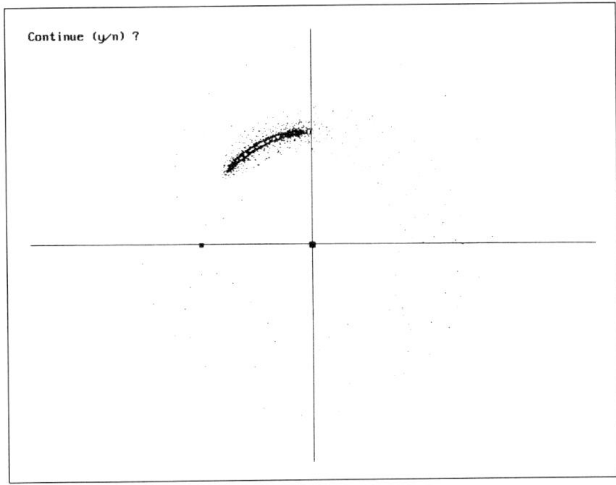

Graphic presentation from THREEBOD.

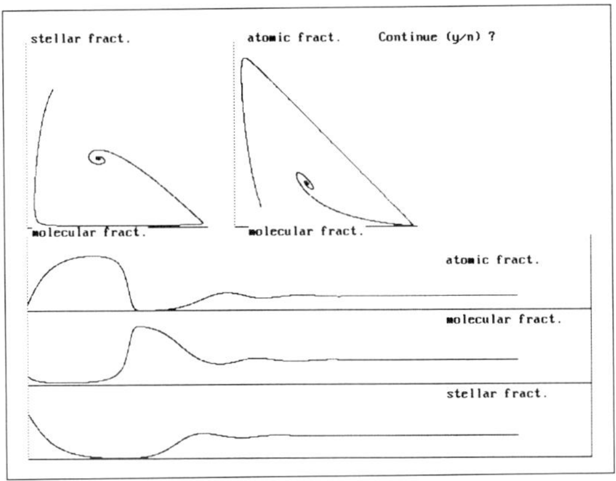

Graphic presentation from STARMOD.

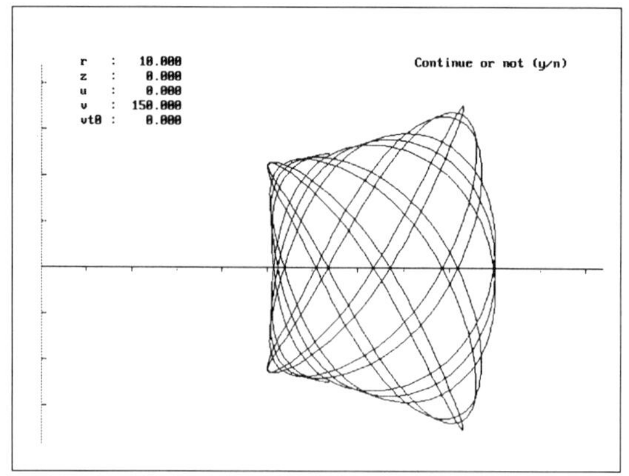

Graphic presentation from `GALAXY`.

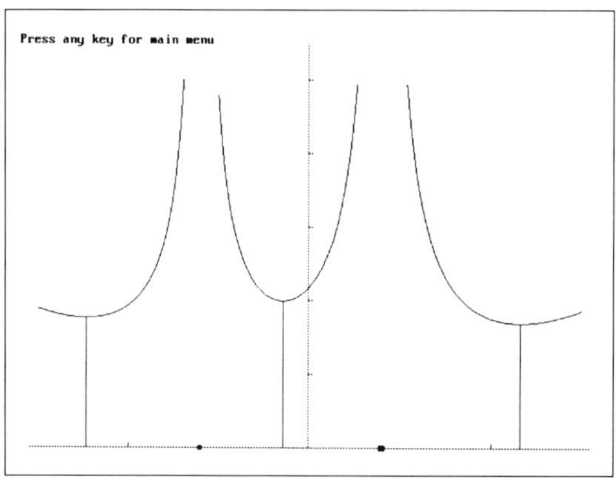

A graphic variation from `EQUIPOT`.

Order Form

The computer listings presented in this book are available as listed below (compiled, executable files including graphical representation of the results are also included.) Please send me the following magnetic media versions of the programs. (Note: This software is sold only to purchasers of the book):

○ **QuickBasic 4.5 for the IBM-PC** at $24.95 each. $__________
○ 5.25-inch 360K ○ 3.5-inch 720K

○ **Turbo Pascal 6.0 for the IBM-PC** at $24.95 each. $__________
○ 5.25-inch 360K ○ 3.5-inch 720K

Handling[1] **1.00**

TOTAL $__________

I wish to pay with:
○ **Check** ○ **Money Order**
○ **Visa** ○ **MasterCard** ○ **American Express**

Card No.______________________________

Card expiration date______________________________

Signature______________________________

Name (Please Print)______________________________

Street______________________________

City, State, ZIP______________________________

Willmann–Bell, Inc.
P.O. Box 35025
Richmond, Virginia 23235
Voice (804) 320-7016 FAX (804) 272-5920

Prices Subject To Change Without Notice

This Book's Serial Number is 087430

[1]Foreign orders: shipping charges are additional. Write for proforma invoice which details your exact costs for various shipping options.